Nuclear War Survival Skills

How to Survive Guide With Self-help Instructions

(Survial Kits for You to Survive Any Atomic & Nuclear Bomb Blast)

Brennan Mitchell

Published By **Andrew Zen**

Brennan Mitchell

All Rights Reserved

Nuclear War Survival Skills: How to Survive Guide With Self-help Instructions (Survial Kits for You to Survive Any Atomic & Nuclear Bomb Blast)

ISBN 978-1-77485-906-3

No part of this guidebook shall be reproduced in any form without permission in writing from the publisher except in the case of brief quotations embodied in critical articles or reviews.

Legal & Disclaimer

The information contained in this ebook is not designed to replace or take the place of any form of medicine or professional medical advice. The information in this ebook has been provided for educational & entertainment purposes only.

The information contained in this book has been compiled from sources deemed reliable, and it is accurate to the best of the Author's knowledge; however, the Author cannot guarantee its accuracy and validity and cannot be held liable for any errors or omissions. Changes are periodically made to this book. You must consult your doctor or get professional medical advice before using any of the suggested remedies, techniques, or information in this book.

Upon using the information contained in this book, you agree to hold harmless the Author from and against any damages, costs, and expenses, including any legal fees potentially resulting from the application of any of the information provided by this guide. This disclaimer applies to any damages or injury caused by the use and application, whether directly or indirectly, of any advice or information presented, whether for breach of contract, tort, negligence, personal injury, criminal intent, or under any other cause of action.

You agree to accept all risks of using the information presented inside this book. You need to consult a professional medical practitioner in order to ensure you are both able and healthy enough to participate in this program.

TABLE OF CONTENTS

Introduction 1

Chapter 1: Surviving An Attack Without Being Prepared 4

Chapter 2: The Plan To Survive 14

Chapter 3: Stocking Your Nuclear Survival Shelter 26

Chapter 4: Historical Background 35

Chapter 5: Catastrophe Awaiting To Take Place 47

Chapter 6: Be Mentally Ready 82

Chapter 7: Ultimate Want: Shelter In Isolation 124

Chapter 8: The Food And Water 153

Chapter 9: The Light And Medicines Chapter 174

Conclusion 183

Introduction

It is undisputed that nuclear war could be a possibility. This has been the case since the time that the nuclear bomb was first invented. Naturally, the moment that an nuclear bomb fell upon Hiroshima and Nagasaki it was a symbol of the power of America. US that effectively stopped the war against Japan.

These two atomic bombs resulted from more than 10 years of research conducted done by European and US scientists. They also set off an arms race that has not been stopped.

In the current world this atomic bomb has been replaced with Fusion bombs. They are built on the same principles like an atomic bomb however, they're much more potent. Any country that has nuclear weapons should be treated with seriousness.

But despite the lessons learnt from the initial arms race and cold war we are, yet again, in the midst of a nuclear war that could cause the destruction of millions of lives. North Korea is the biggest threat, and possibly the most likely nation on earth to trigger an unending war. However it is important to note that it is important to note that the US has not been ruled out of response with nuclear weapons, or even beginning what is known as an "anti-war.

A nuclear-powered missile launched out of North Korea would take approximately 30 minutes to strike in the US mainland. Of course it is possible that US missile defenses US missile defenses could most likely be successful in bringing down a missile.

However, what if the missiles were numerous launched and this missile defense mechanism be capable take on the challenge? In addition, once North Korea has launched, it is very likely that the US would strike against them, while making use of its missile defense.

The result will almost certainly lead to devastation of North Korea; but it could also result in a massive number of deaths, injuries and places that would make it impossible for humans to exist for a long time to come.

Unfortunately, there is very little you can personally do to stop a nuclear strike taking place, especially in the event of an outside power that is the first to launch. Even though nukes have for a long time been used to stop a global conflict, the ambitions of Kim Jong-un as well as Donald Trump suggest that this is about to change.

If you are given a 30-minute warning, you're probably likely to be unable to build a shelter or store food items. It is also highly unlikely that you will be able of escaping the blast zone, particularly near the center of the area of. Even if you're further away, the fear and the subsequent traffic gridlocks make it very unlikely that you'll be able to make it to escape in the time you've got.

It is therefore essential to get started now. It is essential to be prepared, have the required supplies, and a shelter for survival. Use the tips and guides from this guide to make sure that you're able to protect your family members and withstand a nuclear attack.

Chapter 1: Surviving An Attack Without Being Prepared

In many ways, the warning has been issued and the US president is competent and prepared to unleash nuclear weapons. There is little uncertainty that North Korea also has the capability and the willingness to launch nuclear missiles, however, it could mean that the end of North Korea.

If you're a typical person, you're left with two options:

1. Find a shelter for your survival

If there's one nearby your home, then it might be possible to relocate to that area in the near future. It's not a feasible option since you'll have to carry on with your normal life until the attack takes place.

If you don't own the shelter that is to say, you're prepared then you won't be able to stay in it in the event of a disaster occurs.

2. Make sure you wait until the missile is fired

It is the only alternative if you've not yet planned your plan and don't plan to prepare for an attack with nuclear weapons.

If this is the case, then you are able to continue your normal routine until you are alerted You must take action promptly:

Determining Your Location

Finding the most secure place to you will take some time. It's precious time you don't possess when you, at the very least you only have 30 minutes to react.

It is crucial to determine the shelters that are located near your workplace and home This will give you an idea of where to go when the warning is given.

The shelter you choose is any of these:

A public fallout shelter. They may be cramped, but they're designed to handle nuclear explosions and offer you the best chance of surviving the explosion. There will be plenty of companions!

A basement, ideally with walls that are thick. The more deep the basement, the more spacious.

* A building that is old and has strong stone walls is likely to withstand the first blast.

It is crucial to recognize the two primary elements to be considered when dealing with nuclear explosions.

A) the Initial Blast

The initial nuclear explosion will produce a blinding light. it could actually blind you if observe it while less than 50 miles from the site of detonation.

But you shouldn't be standing in the middle of the road to gaze at it! The entire process of explosion takes about ten seconds, after which there will be a rumbling of light that will be visible in the sky.

The temperature at the center that is the source of this explosion comparable to being in the sun's rays and everything that is that is above ground in its immediate area is going to be destroyed.

The explosion can create a shockwave that can travel for many miles, and cause structures to collapse. It can result in internal bleeding.

This is why it is essential to be at least ten miles from the center and in an environment that is more likely to stand up to this extreme heat. It is also safer to be underground.

B) The Radiation

Source: Bruce Blaus; licensed under the Creative Commons Attribution-Share Alike 4.0 International license.

The second component in the explosion is more damaging over the long haul. The nuclear explosion can create a large quantity of radioactive particles to be released from the area of the detonation, and then out into the surrounding region.

If you're in close proximity to this radiation, the high level of radiation could cause death. If you're away from the blast zone, the low doses can result in genetic mutations or cancer as well as a host of other signs.

If you survive the blast, you are at risk of getting radiation poisoning that will impact the remainder the rest of your existence.

This is the reason it is ideal to pick a spot that is as far from the centre that blast is you can and also one which is underground.

Moving during the Explosion

It has been suggested you could shield yourself away from initial explosion, and then move to a more suitable place before the fallout strikes.

A study has shown that the efficacy of this approach will be contingent on the length of time you are spending in the first location, compared to the time it takes to travel to another location.

If you're in this position, the truth is that this is the only choice! Be prepared to withstand the shockwave and heat first before dealing with radiation.

A shelter of any kind is better than one when dealing with the initial blast. However, if you're trapped outside, lie on the ground and take cover of your head. This could suffice to withstand the blast wave, depending on how far away from the ground.

Supplies

Credit Squirrel_photos. License by the Pixabay license

If you're caught at work, or you aren't prepared for a nuclear attack, you might be faced with limited items. However, it is vital to carry all of them carry.

You'll need:

* A bag that can keep your things in

* Food that is dried and canned in the amount you're able to carry within the limited time that you have.

* Tablets for purification and water. The average adult requires one gallon of drinking water every day to drink cook, wash and drink. You won't be in a position to carry or grab the

majority of this. The tablets for purification could help when you discover water.

* Candles or torch, If you have them available

* A spare set of clothing or blanket

• A radio portable as well as blankets

Take everything you can within five minutes, provided you can effortlessly carry everything. Make sure to get your refuge.

These steps could protect your life in the event of an actual blast, but getting through the consequences of the explosion is a separate matter altogether.

Credit: Kalhhh licensee by the Pixabay license

The following information will help you realize how fatal the blast was at first:

Five miles away far from blast zones, you are susceptible to third degree thermal burns.

At 20 miles , the temperatures can burn your skin to the point of being completely stripping your skin.

The shockwave emitted by the blast is expected to travel at 600mph, it will level anything that is in its path in the initial 5 to 10 miles following the blast.

Steps following the Blast

After the initial blast is over, you must immediately move to a more secure shelter in case you're not already in one. Every minute you spend in the open air will expose you harmful radiation.

After you have moved into your home, you must take the following steps as much as you can:

Take off your clothing when you were out in the open during the blast or moved your location. They'll be covered by radiation, which will penetrate the skin. Seal the bag, preferably removing the shelter. But, don't continue to open and shut the doors to your shelter, as it will allow more radiation into your home.

• Wash your body to get rid of radioactive particles. If you can taking a shower with hot water is an excellent idea. However, it is best to be careful not to scratch or rub your skin since this can stimulate radiation to your body.

Although you may apply soap to your body, and even shampoo, it's not advised to apply conditioner if you are a victim of. This is due to the fact that it can effectively bind the radioactive particles to your hair, hindering you from getting rid of it quickly.

• Set your radio up and check if you are able to create an audio signal. It is not a good idea to drain your battery by allowing this to remain on. Instead, set it to a live radio station and switch it on each half an hour for a couple of minutes. This will keep you in the loop with the most recent developments.

* Be in the shelter. So long as you have water and food, do not go outside the building. This will shield you from radiation. The longer you stay in the body, the less it will be damaged by radiation that is released by the nuclear explosion.

It is important to note that within two weeks , the radiation levels in the air will be less than one percent of the amount it was when it occurred the explosion. It's safe to travel within and prepare the next steps.

Being prepared for the worst is contingent on finding a suitable refuge and being armed with enough provisions to last at minimum two weeks. There is a chance for emergency personnel to arrive to rescue you. However, this might not be the best alternative if they don't be aware of your presence or lack the resources to search the area.

It is very likely that there is multiple missile strikes and the government will decide on their actions based on the needs being done in order to keep the officials in good health and running the country. Be patient and be your best partner in this situation.

Chapter 2: The Plan To Survive

In this moment, you can be fairly certain that you'll be able to be able to survive the nuclear explosion, provided you're not far away from ground zero. However, this can be deceiving the most challenging part of survival is following the explosion.

The first step is to be patient and wait for the radiation to pass. The fact that you have enough supplies to last two weeks in the radiation chamber will reduce the effects of radiation on your. When the two weeks have passed, unless the radio informs you otherwise, you will be safe to move about. However, it could be months, weeks or even years before the houses are rebuilt and the infrastructure is repaired.

Assistance will be contingent on the amount of attacks that occur and the importance accorded to your specific area.

If you prepare for an attack early, you significantly increase the odds of being able to survive. Additionally, you'll have the ability to decide whether to remain in the area following the explosion or try to move to a different area.

These steps are vital in planning for survival:

The Plan

The first step is to formulate your action plan. It is the actions you must do when you receive an alert for nuclear danger. In most cases, you'll be working as well as at your home. It is important to consider the distances involved, and then calculate the location you will be able to reach, regardless of whether it's separate from your family.

This is the main motive behind the plan to ensure that each member of your family is aware of which direction to take as well as what they should do. If you're not able to connect, then the plan will specify the time, place and date you'll get together.

Be aware that the best warning you can get will be 30 minutes , but it could be less than that, if you are within ten minutes from your home, then you're unlikely to be home on time.

There is a possibility to try to make it halfway home, take refuge from the initial blast , and continue your journey until the radiation strikes, but it could be an extremely risky choice.

Your Shelter

It is now the perfect moment to start thinking about building your own home. The most convenient place to go to when you've got an alert and are in home is in the lower levels of your home or the basement. The best location to build your shelter.

You must also determine if there are any shelters for public use close to your school or workplace It will ensure that you know where every person in your household is.

The shelter should be constructed in accordance with these guidelines:

Underground is the most effective. The more dirt and walls between you and radiation, the less likely it is that it will be harmful to you.

It should be large enough to accommodate your entire family, with enough room for sleeping arrangements as well as cooking, toilets and laundry;

* You'll require an area for your supplies, and be ready to store it!

* Although steel and lead are the most effective options for a fallout shelter, the most simple one to find is concrete. Its 2 1/2 feet thick , it will give you a solid shield from radiation.

The most effective shelter is one that is built into the ground and that has at least an earthen wall of three feet around it, with two and a half feet of concrete roof.

You could choose to dig a big hole in your garden to build this kind of shelter. This isn't an option to wait until the last minute; it is time to begin building right now!

The hole should be constructed with a concrete foundation and brick walls to prevent the soil packed with dirt from getting inside your home and provide an additional layer of security.

It is also important to take into consideration the accessibility. It should be through stairs, which are separate by the building using an entrance and you'll require an airtight entry point to the outside.

A shelter built from scratch can also let you to create rooms within the shelter. A separate bathroom, sleeping space, and living space can ensure that your stay is more comfortable. It is recommended to sketch out the layout prior to digging.

It is also important to determine if you require permission to construct an structure similar to this in your backyard.

There are many aspects that you must consider when creating your home:

Ventilation. The more airtight your shelter is, the more likely you will not be afflicted by radiation exposure. But, you'll need some airflow to ensure that you don't suffer from a lack of oxygen!

It is recommended to install an intake tube for air that has an excellent radiation filter to ensure that radiation is not taken into the home. It is possible to use a fan to draw in air as well as a secondary pipe equipped with a valve that will stop the flow of air and allow the air to be drained to exit the house.

It is important to remember that radiation flows in straight lines and towards the direction of the wind. Making 90 degree turns on the pipes and doors can help stop it from getting into your home.

Power - It's likely to be that power lines will withstand the explosion. You'll need to depend on a generator, and even solar panels are susceptible to damage due to the heat generated by the blast.

This means that you will require an additional space for your generator , with its own outlet and inlet pipes, and a sufficient quantity of gasoline. You should try to soundproof this space as the sound of the generator running can quickly become irritating within a room that is enclosed.

Toilet - It's advisable to purchase portable toilets that do not require water. You do not wish to waste your valuable resource by flushing your toilet. Odors can be diminished through the use of litter for cats.

Secondary Hatch - You'll require a second entry or exit. This is crucial when your main entryway is blocked or damaged. The second entry point can be easily concealed with an opening that opens to the inside and sanding it to the exterior.

It is recommended to mark the area at the exit secondary to the main one using the help of a few plants or some other kind of structure, but don't put them over the hatch!

Heating and Food preparation – Your home could be warm enough because it is underground , and the body heat generated by four members of your family could suffice. But, if it isn't, you'll need to think about cooking methods and heat the house. It must be a clean burning fuel to reduce the danger from Carbon monoxide poisoning.

Washing Facilities There must be two wash facilities prior to entering the shelter. This allows you to wash all clothing that has been in contact with radiation.

The other should be within your shelter, ready to be used as you wait out the radiation.

Of course, building such a shelter will require a significant hole and it's likely that you won't succeed in achieving this without neighbors being aware. This could be a problem in the event of a nuclear attack because they might want to come to the shelter, and you would not be able to accommodate additional individuals.

It is for this reason that you may decide to build any of the following shelters:

Basement Shelter

If your home is built with an underground space, it's possible to build a room in it, which can provide an adequate amount of protection against the radiation blast as well as radioactivity.

This method has the benefit of being able to be completed without attracting attention to the work you're creating.

The most effective method is to construct a brand new wall within the existing wall, and made with natural stone or concrete. It should be about 2 feet in length.

In this area, you'll have to follow the same strategy as in the shelter you built in your garden.

Air is among the most important issues to consider in the field of radiation. The air itself isn't radioactive, however it does contain radioactive particles. These particles must be cleaned before they can enter your home.

The Earthbag shelter

They can also be utilized to make the appearance of a space in your basement, or to create a shelter within your backyard. They can be utilized above ground or beneath and can be easily moved by just one person.

If you'd like to keep from digging a hole, you can build shelters out of these earthen bags using an organic rise in the soil.

Source: Patrick Mueller; licensed under the Creative Commons Attribution-Share Alike 2.0 International license.

They can also be utilized similar fashion to bricks. If they're properly compressed, they'll remain extremely solid in all weather circumstances. You could even protect the entire structure with extra soil, if it is above ground. This keeps it from being seen and will can protect you from nuclear attack.

It is vital to keep in mind that you need to install a waterproof lining in order to make sure there isn't water intrusion.

Earthbag shelters are typically in a circular shape since this kind of structure is extremely strong.

Also, the power of ventilation and room dimensions. should be considered thoroughly prior to building the shelter.

Earthbags can be used to build walls for your underground shelter. They are an efficient and inexpensive construction material. They can even be a great method of using the soil that you have dug out from the earth!

Natural Shelter

There is a chance that you have an old cave or mine within your area. It could be a suitable option to build your nuclear shelter because it already provides the best level of security.

It is important to look over the below items:

* There's a decent airflow in the mine or cave, but it's one can be controlled by filters or 90 degree rotations to stop radiation poisoning.

If you're thinking of buying an old mine, you'll be required to confirm that there isn't a gas build-up within the mine.

* Location If the mine or cave are located on your property, this shouldn't be a problem however if it's open to the public, you will not be able alter the location prior to the blast, and

you may discover that a lot of people visit the same location.

Sandbags, also known as earthbags, are a great option to block huge spaces within these structures. They can not only add 90 degree angles to reduce radiation, but you can also use them to build an outside shelter for cleansing prior to you enter your main shelter.

Also, you must examine the water table prior to building any structure in a cave or mine because you don't wish to be inundated when you get there!

Every shelter you construct is recommended to be built underground and have at minimum three feet of soil or concrete on top. This can reduce the risk of exposure to radiation. The main factors to take into consideration are the same regardless of where you are and the kind of structure you're planning to construct.

The most important question that determines the type of shelter you plan to build is whether or not you would like your neighbors to know what you're up to or not.

Chapter 3: Stocking Your Nuclear Survival Shelter

As previously mentioned, as long as you're at five miles away from the point of explosion, it's relatively simple to get through the blast and move into a refuge to protect yourself from radiation that follows.

The trick to survive the next step is to plan sufficiently in advance to ensure to ensure you have everything needed to live for at least two weeks in your home. Also, you should consider the things you'll need in the event of your escape; it's unlikely that you'll be able to return to your normal life.

* Food

You should think about what foods your family eats that can be stored in the for a long time. Food items like rice, pasta beans, honey, and pasta are all able to be stored for long periods of time.

Gloves

Eye wash

Dust masks can assist in preventing your breathing in radioactive particles

Potassium Iodide tablets are best consumed immediately you realize there's a nuclear attack. The tablets can boost your thyroid and decrease the chance of getting cancer in the near future.

Painkillers, like paracetamol or ibuprofen

Aspirin can be helpful to thin blood should it be needed.

Stomach medication

Anti-diarrhea treatments

Prescription medication These medications' dates are required to be checked and rotated.

Glasses for spare

* Batteries and torch

A well-designed shelter should have its own power source that isn't damaged by nuclear explosion. But, it is not guaranteed this , and you won't be able to save fuel by running a generator to keep a light on.

Instead, you should stock up on batteries-powered lamps and torches along with an array of bulbs and batteries. They'll be much more useful when you live in your house.

* Radio

Radios serve two functions Its primary function goal is keeping you informed with the latest developments in the nation or across the globe. You must know the number of people and regions that are affected, and also listen to any government broadcasts that might give you advice regarding emergency rescue.

Ham radios can be a fantastic option to attempt to talk with others who survived. You will require electricity as well as an aerial above ground which should be able to withstand the explosion.

* Games and Books

You might be amazed at the amount of time you'll must fill up while inside your home. There's nothing to do other than sit and wait. A variety of games and books will aid in the process as well as keep the spirits of everyone in check.

* Shampoo & Towels

It is necessary to cleanse however, you should do not rub your face if believe you've received radiation. This can help stop radiation from getting absorbed into your body.

It is also necessary to maintain a healthy environment in your house; something that shouldn't be ignored if you want to be well. Naturally, you'll need towels to dry after you wash.

* Blankets

Heating the nuclear survival shelter isn't an easy job. If you don't have electricity available, it is possible to rely on a stove that burns wood but it will also mean another escape point for the outside world that can allow air to enter.

Blankets are a great method of using your own body heat to keep warm. Having a large stockpile of them is crucial.

* Clothing

If you've had exposure to radiation while walking to your home, you'll have to get rid of and store your clothes to avoid radiation poisoning. Even if this isn't an issue, a few outfit changes can aid in feeling more at ease while waiting for radiation levels to decrease.

* Radiation Meter

The personal radiometer is simple and fairly inexpensive to purchase. It allows you to determine the amount of exposure you are exposed to.

* Plastic Bags

An array of these can assist you to tackle the issue of waste and cat litter can be helpful in eliminating any unpleasant smells that may be present in your shelter.

* Personal ID

The post nuclear war world could appear completely different from the one you were accustomed to. It is therefore important to establish who you are. Having copies of the original ID documents can make this simpler

* Cash

Banks are not likely to operate following the nuclear strike In fact, cash could be of no usage. It is nevertheless important to have some on keep on hand in case of.

* Chargers and computers

They could have limited utility following the event of a nuclear attack and will require a power source to charge the devices. But, your computer likely have important data that you do not want to lose. It may contain a wealth of information on the most effective methods after a nuclear attack and medical emergency.

This can be extremely helpful.

There are additional items that are suitable for inclusion in your survival kit items that will assist you to live more comfortably after and during an attack by nuclear weapons could be included, provided you have enough space.

The most important thing is to make sure you choose items that will bring significant benefit to the family and you.

Being able to survive the explosion as well as the first fallout just the beginning. Your supply of supplies could be required to remain on hand for a long time as you wait for assistance to arrive.

Final Note:

Many survival plans for the end of the world events include having a garden that is ready to be used. While it will be a source of some food sources, it's also likely that the soil will get polluted by radiation.

This will result in your crops becoming affected, as are the animals that are grazing in the vicinity. Food you've gathered inside your shelter is your most secure source of food and the more you've built up a stockpile, the more you'll be able to be able to survive.

Chapter 4: Historical Background

Like the nuclear and atomic bombs, these weapons, which depend on nuclear reactions to produce their explosive power and are extremely powerful. Following World War II, scientists began working on the first real nuclear weapons technology. Two nuclear bombs were dropped on two cities in Japan during Japan and the United States during World War II. After this war both the Americans as well as the Russians took part in a worldwide nuclear arms race during the Cold War, a period of nuclear expansion. A few German scientists created the first possible atomic weapon after they found nuclear fission in 1939 at an Berlin laboratory for science. The energy is released as a radioactive atom splits into smaller particles. To develop nuclear weapons and the associated technology discovering nuclear fission proved to be an important step towards the proper direction. The power of nuclear bombs is generated by processes of fission. Thermonuclear weapons, commonly referred to in the field of hydrogen bombs produce their destructive power through nuclear fission and the process of fusion. In the event that two lighter elements fusion in a process of nuclear fusion the energy released is then released.

The Manhattan Project Manhattan Project

A team led by Americans was working using the pseudonym "Manhattan Project" to develop a functioning nuclear weapon during World War II. From during the 1930s, there was a suspicion of German scientists were working on nuclear weapons. This was the time the time that they started the Manhattan Project was kicked off. As per a presidential decree that was issued on December 28th 1942, scientists and military officials who were involved in nuclear research could be part of the Manhattan Project. Manhattan Project theoretical physics team head J. Robert Oppenheimer commanded the Los Alamos, New Mexico part that was part of the Manhattan Project. It was the Trinity Test, the first nuclear bomb that was successfully detonated, occurred on the 16th of July, 1945 in the remote desert close to Alamogordo, New

Mexico. The era of nuclear weapons began with the detonation of a

massive mushroom cloud that reaches 35,000 feet.

Hiroshima and Nagasaki Bombings

In 1945, the scientists at Los Alamos had created two different kinds of nuclear bombs "the Little Boy," made of uranium as well as "the Fat Man," made of plutonium. Even after the conclusion of the European conflict in April, the war with Japanese and American forces within the Pacific continued. In the latter part of July The Potsdam Declaration, issued by President Harry Truman, urged Japan to surrender. In the aftermath of the "Little Boy's" blast the city was reduced to rubble and 80,000 people were immediately killed. Over a 10th one million people perished because of radiation poisoning. The first time he struck, "Fat Man" was said to have killed more than 40,000 victims. The second blast was not aimed at Nagasaki. An attack by firebombs on Kokura the largest Japanese munitions plant, stopped American bombers from targeting the city's munitions facility. The American jets then switched their focus towards Nagasaki as their third and final target.

The Cold War Cold War

Through an international network of espionage, the USSR came up with ideas for a fission-type weapon and quickly discovered uranium sources throughout Eastern Europe. Soviet scientists launched the first weapon made of nuclear material in August 29th, 1949. The US immediately responded with the process of creating more advanced thermonuclear weapons in the year 1950. It was the time when Cold War weapons race had begun and many nations, including the Americans as well as the Russians have made nuclear research and testing prioritizing nuclear research and testing as their top priorities.

Nuclear Non-Proliferation Treaty

The negotiations to limit the proliferation of nuclear weapons during 1968 were led by the United States and the Soviet Union as well as the rest of the world was required to follow in their footsteps. NPT (also called the Non-Proliferation Treaty or NPT) was signed on the 1st of January in 1970. States with nuclear weapons and states that did not have nuclear weapons were classified into two categories towards the end of Cold War. Since the 1950s the nuclear arsenal was known to be in the possession of all the mentioned nations. Nuclear-armed states agreed to not to deploy their arsenals against states that are not nuclear or aid them in the acquisition of nuclear weapons. Their arsenals of nuclear weapons were set to be reduced in time toward complete elimination. Belarus, Kazakhstan, and Ukraine were the sites of many weapons. All of the weapons were given back to Russia and removed from service.

Nuclear Warfare

Nuclear weapons are used in a conflict or in a political strategy is known as nuclear war (also called thermonuclear war or atomic warfare).

It is possible for nuclear weapons to produce an extended radiological impact and thus, they are considered to be explosive weapons, in contrast to conventional weapons. The long-lasting effects of massive nuclear exchanges, such as the potential for the possibility of a "nuclear winter" lasting for centuries, decades or even millennia following the first nuclear strike, could be devastating. Even if millions living in rural areas died during the event of a nuclear winter, according to certain experts, the vast majority could live. There are some who believe that the repercussions of a nuclear holocaust like nuclear hunger or the breakdown of society could result in the death of nearly every human being in the world.

There have been only two occasions where atomic bombs have been employed. The first was in August of 1945. A weapon made of uranium named "Little Boy" was blown up over Hiroshima, Japan. A plutonium-based implosion weapon as its primary component (codenamed "Fat Man") struck Nagasaki, the Japanese town of Nagasaki just afterward on the 9th of August. Japan's demise was aided by two bombs that led to the deaths of more than 100,000 people. In 1974 and 1998 India and Pakistan two countries that have many years of animosity have acquired nuclear weapons. For instance, it's not known what number of nuclear weapons were constructed by Israel during the 1960s as well as North Korea in 2006. Israel's government has never acknowledged or denied having nuclear weapons, despite the fact that it has constructed the reactor as well as the processing facilities needed for this. In the case of developing nuclear weapons during the 80s South Africa was the first nation to completely eliminate its domestic arsenal and cease manufacturing of nuclear weapons of its own volition. In the interest of testing and demonstration approximately 1800 nuclear explosives been detonated. When there is a conflict that involves an use of nuclear arms there are typically two distinct subgroups with each possessing distinct implications, and the capability to use a variety from nuclear arsenals. The first type is a "limited" conflict with nuclear weapons (sometimes called an attack or exchange) in which nuclear weapons are employed sparingly however, there is a supposition that a nation could use them more frequently at some point in the near future. The small number of nuclear bombs used against the military could be increased with the increase in the quantity of bombs that are used or intensified by choosing different targets. In the event that an opponent makes only a small amount for nuclear war, small attack is considered an acceptable response rather than a complete attack. Nuclear weapons can be used as part of a major assault on the entire nation which includes economic, military and civilian infrastructures. It's obvious that such an attack will destroy the target's economy, societal and military infrastructures as well as the biosphere in general. According to some, an uninvolved battle could eventually turn into a complete nuclear conflict. The global nuclear catastrophe,, as many have described it, is the outcome of a complete nuclear conflict between two superpowers, which would last for decades to unfold and would destroy the world like an "full-scale nuke war" could, but with a longer road towards destruction. In general, the likelihood of a huge nuclear war with the superpowers of nuclear war diminished after the dissolution of the Soviet Union and the end of the Cold War in 1991. There was a shift in the direction of nuclear weapon debates to prevent regional nuclear wars, as well as potential nuclear terrorist attacks.

Nuclear Bomb

In a different way"nuclear weapon" or "nuclear weapon" is a reference to any explosive gadget that can be detonated using the nuclear reaction, no matter if it's fission (in the instance of"fission bombs") "fission bomb") or an amalgamation of nuclear reactions (in the instance of"nuclear weapon") "nuclear bomb" or "nuke") (thermonuclear bomb).

When it comes to uranium-239 and Plutonium-239, the neutron hits the nucleus inside an atom, which causes the nucleus to split in two nuclei that contain about half of the neutrons and protons of the nucleus that was originally created. The atomic bomb is produced by a chain reaction in which fissions rapidly increases, devouring nearly all fissionable substance and creating an explosion. Combining the naturally occurring uranium isotope, uranium-235, creates greater numbers of neutrons in fissions than other isotopes that can be fissionable and are found in a small fraction of part of the naturally found uranium-238 isotope. Plutonium-239 is the same substance like Plutonium-238. These are the main elements that can be fissioned in nuclear bombs. Subcritical the uranium-235 element is not more that 0.45 kg (1 pounds) due to the fact that neutrons produced in fission are not likely to hit another nucleus, causing the fission process, thus which prevents chain reactions.

A fissionable assembly of materials must be moved from a subcritical to critical status rapidly in real life. If two masses that are subcritically different joined and their mass is subsequently critical. There are a myriad of ways this can be achieved. Utilizing powerful high explosives you can launch the subcritical round of a fissionable material in a ring inside hollow tubes. In an implosion-type nuclear weapon fissionable central core, it is enclosed by high explosives which simultaneously explode, which compress the fissionable substance under enormous pressures to form a dense mass that is rapidly reaching critical. Beryllium oxide or another substance could be coated with the fissionable material , and reflect neutrons back into the fissionable materials, where they could cause further fissions. This is an important aid to reach criticality. Fusion materials such as deuterium and tritium can also be employed for "boosted fission" devices. The explosions of fission can be enhanced by the materials ability to combine with other materials. The amount of energy released through fission is enormous in comparison to the volume that is involved. When uranium-235 has been fully fissioned release the exact amount of energy as 16000 tons of TNT. In the event that an atomic weapon goes off massive amounts of energy heat are released, and the weapon reaching temperatures that exceed a million degrees. In the wake of the heat, a huge fireball forms which could ignite the ground to cause destruction to a large city. In the event of an atomic explosion convection currents pull dirt and debris that has fallen from the ground to the fireball, creating the cloud that resembles a mushroom. After the bomb explodes an intense shock wave immediately begins to form and propagates for a long distance before disappearing.

As the blast progresses the radiation that causes death is absorbed quickly, ranging from 1.5 up to three kilometers (1 up to two miles) from the blast. The winds scatter tiny particles of radioactive material that have vaporized from the fireball that is located in the stratosphere or troposphere and this is referred to in the form of radioactive particles. Strontium-90 and plutonium-239 are the two radioactive pollutants that have a long life. A small amount of fallout exposure during the initial few weeks following the explosion could cause death, and even just a tiny amount increases the risk of developing cancer.

Chapter 5: Catastrophe Awaiting To Take Place

Humanitarian Impacts and Risks of Nuclear Weapons Deployment

CRC and IFRC organized an all-day discussion with experts 2 March 2020 regarding the humanitarian consequences and threats of nuclear weapons. The event was organized through the International Community of the Red Cross. The participants were asked to create an assessment of the nuclear weapon usage and the effects of testing on both the natural and human surroundings. More than 45 countries as well as UN agencies as well as civil society groups attended the event including experts who came from Sciences Po, the Committee of Scientists as well as The UN Institute for Disarmament research's Gender and Radiation Impact Project. According to the study, that ICRCS published the findings of the meeting, they were condensed. This conclusion may or may have been viewed as untrue by participants. A lasting impression has been left by doctors from Japan's Red Cross and ICRC, who witnessed the destruction in 1945, and also the ICRC's efforts in eliminating nuclear weapons in the past 75 years. It was the International Community of the Red Cross as well as other organizations began recording the environmental, physical and medical impacts of the nuclear bombs dropped upon Japan in 1945.

Since the initial nuclear weapon was used and tested, they have been the subject of research investigation. In 1987 in 1987, the World Health Organization published seminal research into the effects of nuclear explosions on the health of people and their health. While nuclear explosions can release radiation and radioactive fallouts scientists discovered that the current health systems aren't equipped to deal with the consequences either in the short or long-term. While nuclear weapon use and tests have been linked with long-term human effects since the 1970s, the evidence and capacity that international organisations have to aid the affected population has increased. Recent studies have examined the effects of radiation ionization to the wellbeing of both men and women in different ways from those of other and also the environmental impacts that can be long-term from testing nuclear weapons, such as infant mortality rates and mortality rates and the global consequences of a nuclear conflict. Despite the fact that a lot of the environmental and health effects of nuclear weapon use and testing are not fully understood new and relevant data has emerged from the latest research.

A greater amount of research and speedier studies are required to comprehend the long-lasting effects of nuclear weapon testing. The ionizing radiation effects from tests conducted in the past are being felt by people living in areas like the Marshall Islands, Kazakhstan, Algeria and in the United States of America. Residents of regions that are radioactively contaminated complain that they do not receive enough information about the dangers of living in an ionized area , and the effects of radiation exposure on the next generations. It is crucial that future research studies tackle issues of intransparency and dislike of the opinions of lifestyles, values, and expectations expressed by various segments of society. Ionizing radiation has been proven to affect women and children more than males, however there is not much information on the effects of radiation on fertility. Here are some possible areas for research in this area:

* Biologically speaking, sex is at risk because of radiation.

* Radiation-related damage has the highest disparities between girls and boys when it is administered to infants as well as toddlers.

* The illness could be caused by radiation.

Risks

A risk assessment for nuclear weapons must be accompanied by evidence of the probable effects of a nuclear explosion. While nuclear weapons haven't been used in armed conflict from 1945 onward, there were numerous close encounters. They were utilized accidentally due to mistakes or miscalculations. The conferences held between 2013 and 2014 were focused on the human consequences of nuclear weapon detonations and it became clear that these risks do not result only from human error, but also from the deliberate design of nuclear weapons.

Human errors and cyberattacks can cause damage to nuclear military command and control systems.

The nuclear arsenal is deployed on the move including thousands of rockets prepared to launch at any time.

* Concerns over the possibility of non-state actors having access nuclear weapons as well as the associated materials.

To assess the risk of technological advances, it's crucial to evaluate the technologies in conjunction. The processes for making decisions can be altered in surprising ways because of the interplay and interdependence between new technologies. Relying too heavily on technology can result in a misguided belief in the accuracy of data generated by these technologies and could result in unfounded confidence. The incorrect interpretation of the actions of another state could be made worse by introducing or using technology that is new. To be precise, providing an objective and accurate analysis of these risks might not be possible and may cause overconfidence in the ability of a person to make such a decision. The probabilities of nuclear catastrophes are based on past nuclear disasters and don't take into account new routes that are not yet explored. Employing language that is based on risk could make an appearance of control and control by suggesting that all possible paths to disaster have been identified and taken into account. The potential of nuclear weapons for usage can be understood better by utilizing the words "luck" or "vulnerability" are employed to refer to our inability to manage and control the use of these weapons.

The study of the immediate and lasting effects is essential as it allows us to understand the specific characteristics that these guns possess. In addition, it is the basis for planning and responding to emergencies and also protecting the rights of those affected. International legal system for humanitarians (IHL) requires evidence of the human rights consequences to decide if they are suitable for use. It's a starting to discussions on nukes, disarmament as well as non-proliferation. As the likelihood of the use of nuclear weapons rises the evidence of the negative impact on human health and the natural environment is more vital. Initiatives to decrease the danger of using nuclear weapons should be lauded by human rights activists. Whatever the price nukes should be avoided at all cost. The legal obligations of governments in pursuit of nuclear disarmament like the ones in the Treaty on Non-Proliferation of nuclear weapons, are not able to be replaced by reducing nuclear risk. Once nuclear weapons are banned and eliminated, they can never again be used. But, more research is required into the human and environmental effects that nuclear weaponry has on the environment. A deeper examination of longer-term environmental and human rights implications of nuclear weapon testing and the age- and sex-specific consequences of ionizing radiation and the possibility of transgenerational transmission is required.

Effects of Nuclear Explosion

Nuclear weapons may be fired simultaneously, leading to an international catastrophe and radioactive fallout. A large portion of Earth could be deemed inhabitable because of nuclear conflict. Likewise, civilizations could fall apart which could lead to the destruction of the human race and/or the end the entire lifeform on Earth. The occurrence of nuclear winters, firestorms extensive radiation illnesses or electromagnetic pulses could be the result of the aftermath of nuclear war, and the devastating destruction of cities due to nuclear explosions. Researchers such as Alan Robock have theorized that the thermonuclear conflict could result in the end of modern civilizations on Earth because of a lengthy nuclear winter. According to one scenario Earth's temperature could drop by about 7-8 degC (13-15degF) following an entire thermonuclear conflict.

Based on early Cold War research, a global thermonuclear conflict could result in billions of people living through the immediate consequences of nuclear weapons and radiation. Based on the International Physicians for the Prevention of Nuclear War the secondary effects of nuclear war, like destruction of the environment as well as social destabilization as well as economic decline, could cause the extinction of humans. India as well as Pakistan are predicted to be able of producing a nuclear winter, and killing more than a billion people, with just the power of 100 Hiroshima (15-kiloton) nuclear weapons. The risk of nuclear catastrophes heavily affects people's views regarding nuclear arsenals. Mutually Assured Detruction (MAD) can be described as a standard scenario in survivalism , and is a component of the concept of security. This concept of nuclear destruction frequently appears in science fiction works dystopian, post-apocalyptic and dystopian literature and in cinema.

Effects on the Short-term

Two immediate results from a nuclear explosion which are gamma radiation and neutron radiation. The nuclear reactions that the weapon produces generate this direct radiation lasting less than one second. A 10-kiloton explosion emits almost 1 mile of lethal natural radiation. However, direct radiation has little significance in the majority of weapons because other harmful effects are often spread over larger distances. Neutron bombs are a distinct one, as they boost radiation but minimize other harmful effects. Exploding nuclear weapons explode within a matter of minutes. The solid and cold substance has transformed into more hot gas than the core that is the Sun. The gas that is heated emits the X-rays which reflect their energy back into the atmosphere around it, warming it. In just 10 seconds following the explosion of a single megaton the superheated air is able to reach the size of a mile. The first phases of a single-megaton gun are much larger than that of the Sun up to 50 miles due to the heat generated by the fireball. The bright star emits warmth in addition to light.

The weapon's explosive force is greater than a third the result of its thermal flash which can last for a lengthy duration. It can be as far as 20 miles from a huge thermonuclear blast, the intense temperature can cause flames to ignite and cause severe burns to the skin. Flash burns were observed on two-thirds of the injured Hiroshima survivors. The radiations from the nuclear explosion however are so powerful that combustible materials can be lit up with a minimum effort. The rapid explosion of the fireball triggers an abrupt rise in air pressure within the vicinity. The speed of the fireball increases to hundreds of miles per hour the blast wave then decreases in speed. A significant portion of the explosive power and the majority of the physical damage is absorbed through this part. The typical atmosphere's pressure is around 15psi (psi). This is a force of 15 pounds across every inch in your body or your home. Because air pressure is typically evenly distributed across all directions it's unlikely that you'll notice 15 pounds that are pulling on the same sq. inch since fifteen pounds push in the opposite direction. Overpressure, which is caused due to a variation in the pressure of air on both sides an object will be what you feel. The feeling of pressure can be felt in the event that you've attempted the opening of a doors presence of a powerful wind. A pressure of 1/100 psi could make it nearly impossible to open the door. In other words it's because a room contains around three hundred square inches. Also, a tenth of one pound is quite a bit of weight. Overpressures as high as a few psi could be created several miles away from the site of a nuclear blast by blast waves. It's an excellent idea! Even at just one psi of overpressure one's front walls of a small home is 50, 000 square inches of surface area, which is equivalent to 40000 pounds, which is 20 tons, of force. A majority of homes can withstand pressures up to 5 psi without destroying them. Commercial and industrial structures are destroyed at ten pounds of overpressure. Even reinforced concrete structures can be in a state of equilibrium at 20 psi pressure. People who are stressed out, however will not allow it.

As a result, many suffer fatal injuries because of the blast's effects. The place of the explosion is determined by what weapon is used to detonate. When a bomb is detonated thousands of feet up in the sky, it can cause the most severe destruction to structures. The destructive power of an airburst is heightened because of reflections of blast from the earth. A ground burst leaves a huge impact and destroys everything that is in the path of its travel, however the explosive power of an airburst is restricted in their size. Cities that are hit by nuclear weapons are likely to use air blasts, while the military targets, like underground missile silos will be more likely to employ ground bursts. The radioactive fallout of both types of explosions differ and will be discussed in a minute. What is the power of an weapon's destructive force? Based on the power of the explosive the devastation radius can be reduced or increased. The weapon's yield is a measure of the extent of space it will eliminate at a given level of destruction. The destructive radius increases in proportion to the cube root of the size of the object since the volume is ininverse proportion to the radius of cubed. The devastation radius can only be multiplied by two and a half times for each 10 times the growth in output. It's more efficient, but not in a manner which is in line with the power. Smaller weapons are far more efficient than one larger one because of the constant increase in destruction as the output increases. For instance twenty 50-kiloton bombs could cause destruction to an area that is three times that of a single megaton weapon. The amount of destruction you wish to cause directly affects the magnitude of the radius that you want to destroy.

The destructive radius is defined as the length of time where the pressure is reduced to around five psi. A lot of people within this range are likely to be killed, however, certain individuals would be left unharmed. If everyone in the 5-psi circle were killed and the outside world was confined in the same way, the outcome is similar to the one described above. In the case of a hypothetical explosion the diagram below shows the way that the zone of destruction changes in proportion to the explosions. In-depth simulations that take the various geographic and environmental aspects are required to precisely estimate the effect that nuclear bombs have on huge populations. The blast waves will disperse within a matter of seconds but the immediate repercussions could remain. If the pressure and heat of the blast continue to die, the flames they cause could result in an enormous firestorm capable of creating its own wind and spread the flames. The hot gases from the firestorm dissipate and are and are replaced by miles of air that slams across the surface. The flame consumes enough oxygen to kill anyone who is left alive as fire and wind intensify the destruction caused by the explosion.

Nuclear Fallout

Nuclear and conventional weapons can cause devastating blast effects. However, the magnitude of these impacts differ significantly. Nuclear weapons are, however generate radioactive fallout. In addition, neutron capture as well as other nuclear processes add materials radioactive to falling out fission products constitute the bulk of the radioactive debris. Isotopes that have a half-life longer than the time of the explosion or other brief effects are known as fallout. The fallout contamination can last for decades, even though death-like effects can last from just a few days or couple of weeks. The current civil defense guidelines recommend that those affected stay indoors up to 48 hours after the radiation levels decrease. In large part, the fallout produced by nuclear explosions is determined by the weapon's type as well as the explosive's output and detonation site. Although it produces high levels of direct radiation however, the neutron bomb is in essence one of fusion devices that produce minimal waste from fission. The fallout of small fission bombs like the ones used for Hiroshima and Nagasaki are common. However, today's thermonuclear weapons employ a fission-fusion-fission architecture that adds a new phenomenon: global fallout. The fission of U-238's outer jacket surrounding the fuel is the main cause of the fallout.

Due to the enormous quantity of radioactive material as well as the time required to allow the cloud of radioactive material to rise into the stratosphere, these massive weapons are able to have an impact across the globe. In the years since they dropped the Hiroshima as well as the Nagasaki bombs, we've not been through a nuclear war however, the fallout from these bombs is one of the results of the weapons we're familiar with. Tests for nuclear energy within the atmosphere was conducted in the years prior to 1962's Partial Test Ban Treaty, and the remains of the radiation have been observed all over the globe until today. If a weapon falls from the sky or exploded at ground level, effects are very different. Since an airburst does not have contact with the ground, radiation can rise into the stratosphere. This minimizes local fallout, while increasing global fallout. Many thousands of tons of dirt, gravel and other debris are hurled into the rising cloud via an enormous explosion of the ground. Rain can wash away huge amounts of radioactive material creating localized regions of high radioactivity. The residents of Albany, New York, were exposed to radiation levels that were ten times more than the normal daily dose of radiation within the area. Although wind speed and direction greatly affect the exact location of fallout The danger zone of deadly fallout could stretch for hundreds of miles downstream from an explosion. Even though isotopes with short lives decay rapidly, the fatality of fallout remains a crucial factor to consider.

Electromagnetic Pulse

There aren't any explosions or local fallout results when an nuclear weapon explodes at an altitude of high. However, powerful gamma radiations could cause electrons to be ripped from the atoms of the air surrounding and the impact could be felt for hundreds of miles if the explosion takes place at a higher altitude. Electrons that are generated by an electromagnetic pulse move around in the earth's electromagnetic field (EMP). An electromagnetic pulse strong enough to damage computers, communication networks and other electronic equipment could be able to affect the entire nation should a single, massive weapon were to strike 200 miles above all of the Central United States. Satellites used to communicate with military personnel as well as surveillance and attack warnings could as well be affected. This means that the military's dependence on advanced electronic devices puts them at risk for the EMP phenomenon. A American nuclear bomb went off into the Pacific Ocean in 1962, 250 miles above Johnston Island.

The crimson aurora which appeared on the sky at night as a result of the explosion was visible in the distance of Australia as well as New Zealand. A stunning flash that was followed by an uninspiring green sky, and the loss to hundreds of road lighting, was observed by Hawaiians located just 800 miles from their island. It's still unclear how these conclusions could be applied to today's increasingly sensitive and widespread technology. It's been difficult to research EMP impacts directly prior to when the Partial Test Ban Treaty in 1963. However, sophisticated systems have been developed to mimic those effects that were caused by nuclear weapons. Electronic systems are "hardened" to guard from electromagnetic waves (EMP). In the event of war, an EMP attack could destroy the communication capabilities of armed forces or command and control networks. High-powered microwave weapons, though not nuclear currently developing by various governments across the world to generate electromagnetic pulses (EMPs). They employ microwaves to degrade electronic missiles, stop vehicles or explode explosives remotely and to bring down a swarm of drones that are not controlled by a pilot. A foe could confuse the results that are a result of EMP devices with those of nuclear weapons, even being nonlethal since they do not produce a blast wave or boom. Does the use of a directed beam EMP weapon, or the higher altitude explosion from an nuclear bomb that creates EMP an act in war that requires the use of nuclear weapons? Should the nation that has EMPed launch a nuclear strike in the event that their electronic alert systems are not functioning properly in the event of an attack? The damaged communications of EMP make it difficult to make nuclear-related decisions. While the solutions to these questions aren't easy the military planners should have them in their arsenal.

Nuclear Winter

According to some experts the destruction of the environment is the result of hundreds of nuclear explosions that occur during the course of a nuclear conflict. Nuclear explosions have been acknowledged as having devastating effects on human beings however, scientists have overlooked the environmental impacts of these explosions for a long time. Nuclear explosions release massive amounts of nitrogen oxides. These could have reduced stratospheric Ozone the layer that shields living creatures from harmful ultraviolet light that comes from the Sun. Additional research suggests that nuclear explosions could result in a temporary cooling of the atmosphere due to the release of huge amounts of dust into the sky, which would block sunlight from reaching surface of the Earth. The year 1983 was the first when scientists began to think about the nuclear smoke that was created by fireballs, as well as the soot and smoke which resulted from the combustion of plastics and petroleum fuels in cities that were devastated by nuclear radiation. This is why the smoke that comes from such sources has the ability to absorb light more effectively than the smoke of a wood fire.) "Nuclear winter" was the first term utilized in this study. The dark predictions about the effects on the environment that resulted from the nuclear conflict were examined extensively by American as well as Soviet scientists. Scientists believe that nuclear bombs explode are the most likely reason for the nuclear winter. If the fireballs were released they could cause havoc on forests and cities that were in their path. Dust and soot will rise up into the air from these fires , and then float for months prior to returning to Earth or being washed away through the winds and rains. To completely cover all of the Northern Hemisphere from 30deg to 60deg, the powerful west-to east winds will have to carry many millions of tons black smoke as well as soot. The dense black clouds could completely block out sun's beams for several weeks. In a couple of weeks the Earth's temperature could drop from eleven degrees to around 22 deg C (20deg up to 40deg F). Photosynthesis in plants could be affected by low levels of sunlight, lethal frosts, subfreezing temperatures and massive doses of radiation resulting from nuclear fallout. This could lead to the loss of significant amounts of life on earth and in animals. Alongside the devastating loss of medical supplies, food and transport infrastructure, due to the extreme cold and the high levels of radiation that could result in a large number of victims of radiation exposure, illness, and malnutrition. In the end, the threat of nuclear war could drastically reduce the global population to a tiny fraction of the population before. Conflicts between nuclear powers in regional areas could damage the Ozone layer. Utilizing nuclear weapons during a regional war could result in an ozone hole that can affect human health as well as agriculture for at minimum a decade, according to research released in 2008. Soot in the upper stratosphere could absorb heat, causing changes to patterns of winds and bringing in ozone depleting nitrogen oxides. When temperatures as high as they are and nitrogen oxides at the same level and the ozone could be reduced to dangerous levels, as low as the ozone hole in Antarctica at the time of spring every year.

Nuclear famine

The absence of accurate data makes it difficult to determine how many people could suffer from nuclear winter. However, it is probable that widespread starvation (known in the field of Nuclear Famine) will play the biggest part. In the event that India and Pakistan are involved in a nuclear exchange , or when both the United States and Russia deploy just a tiny portion from their arsenals three-quarters of the population could be starving. Numerous studies have shown that climate change caused by nuclear war has had a long-lasting impact on productivity in agriculture. Because of the rising cost of food many million of individuals, mainly in the world's poorest countries will be more vulnerable.

Climatic Impacts

The atmosphere's upper layers could be flooded with toxic substances and dust should there be a major nuclear conflict. Before scientists had a thorough study of the possible consequences the world was already in the nuclear age. The scientists found nothing to be that was reassuring about their findings. Ozone is a rich layer in the upper atmosphere that effectively blocks sun's UV radiations. Ozone is a unique form of oxygen. The ozone layer shields Earth from harmful UV radiation that could otherwise penetrate into the Earth's surface. Research has shown that one nuclear exchange could result in a dramatic increase in ultraviolet exposure if the layer were destroyed. The increased UV radiation could affect marine life and trigger painful sunburns for people. The human immune system could be weaker and skin cancers would become more common because of an increased exposed to UV radiation. In the event of a nuclear exchange the smoke of burning cities would be pumped to the sky in a way that is alarming. In the year Richard Turco and Carl Sagan published their TTAPS publication in 1983 they stunned the world by declaring that 100 warheads from nuclear explosion could cause massive global cooling due to soot in the atmosphere prevented sunlight from reaching the Earth's surface. The human race could become extinct should this idea of nuclear winter is extended to its utmost). It's not the first time scientists have thought about an extinction caused by dust. Scientists believe that the dinosaurs' extinction was due to the warming of the planet caused by air dust from asteroid impacts. In comparison to the current climate models the first research conducted in the winter of nuclear energy used a primitive computer model, which sparked intense debate among experts on atmospheric conditions. Although he was not the principal research team, Sagan agreed to lend his name to the report to help raise awareness. Sagan decided to publish his findings to the general media two months before the scientists were scheduled to release the paper. Instead of helping dispel the idea that nuclear war is possible or that systems for missile defense would keep our United States safe from attack This backfired on Sagan and the alarmist scientists such as Edward Teller ridiculed. The idea of a nuclear winter was "extremely uncertain," Teller labeled Sagan as a "great propaganda expert." Due to the destruction caused by the event by the disaster, the general public rejected the idea about nuclear winter.

But, research on the subject of nuclear winter continued. It is estimated that the United States and Russia, even with their current dwindling arsenals of nuclear weapons, could release more than 100 million tons soot and smoke into the air in the event of a complete nuclear war, according to recent studies that relies on advanced climate simulation models. Nearly as much pollution is generated in America each year as it does in this period of time. In the course of a decade, global temperature only improved by 4 degrees or 8 degrees Celsius, which is less than a drop of. It could be a one or two years of subzero temperatures, and 90% less rainfall in the areas of the world's "breadbasket" agricultural zones. A devastating impact on the world's food supply could be created. It is possible to have devastating impacts on the climate from even the smallest nuclear event.

Incitation to Nuclear War

We've so far studied the effects of one nuclear explosion. It's impossible to anticipate an nuclear war, which could include hundreds or hundreds of explosives. Despite the decades of agreements made to limit the amount of nuclear weapons, a large number remain in the arsenals of the world. They are unlikely to cause mass deaths in the event of a nuclear explosion, but there are a few that could. It's difficult to comprehend the terrible consequences of a nuclear war. It's not uncommon to envision an all-out nuclear war that sees both sides employ their arsenals to inflict destruction on each other. Many, including the authors of your piece consider that nuclear arsenals used by superpowers will eventually lead to this kind of catastrophe. Nuclear strategists have analyzed a number of options other than a complete nuclear war. What is the difference between smaller nuclear fights? What happens if they're restricted to a specific amount?

Low yield, conventional military engagement-style nuclear war could constitute an illustration of a restricted nuclear conflict. Here is an example of a scenario Russian tanks and ground forces launched an attack on the Baltic country in retaliation of the 2014 annexation of Crimea while United States is engaged with domestic concerns. In spite of NATO's use force, Russia's determination is unshakeable. Russia responds by deploying tanks, and then destroying NATO sites which has resulted in the death of hundreds of soldiers. NATO has utilized a dial-a-yield setting of just 300 tonnes equivalent to TNT as a last resort to stop Russia's aggression which NATO will not tolerate. The main aim is to convey a message that informs Russia it has gone overboard. NATO has put in place plans to stop Russian aggressiveness from preventing an all-out war within northern Europe. Within the Pentagon's highest-ranking officials, this issue is being debated. Nuclear weapons with low yields may be used as a demonstration of strength and determination to get the other side to a halt from its aggressive behavior. Negotiations will restart and rational voices prevail when faced with nuclear attack. It is believed that everything will go in accordance with the plan. According to what the term "fog of war" suggests, we're in a position where we are required to make decisions even though we don't have enough information to make those decisions. One possible outcome of this scenario is removal or destruction of communications cables or sensors which could have indicated a lower yield for nuclear weapons. Thus, Russia's security could be at risk and it could decide to launch a massive nuclear strike, killing millions. It is likely that this could not be stopped in the event of the possibility of a nuclear conflict. There were no more deaths from hunger due to the devastation caused by nuclear weapons that killed more than 500 million people during the initial strike. It is possible that a small nuclear strike isn't enough? There's a particular reason behind this and it is related to how the Soviet Union team's response to an enacted U.S. nuke attack: the complete nuclear strike on their own. If North Korea was invaded, what would occur? In 2017, some Trump administration officials suggested the use of a "bloody nose" method of dealing in dealing with North Korea. In response in response to the provocative behaviour in they suggested the United States would destroy a vast area. A conventional attack or a low yield nuclear bomb could be a possibility under this situation. Any action from either the United States or its allies could provoke an "all or none" action by North Korea, including nuclear and conventional weapons of massive destruction. But, that's not always the situation. An United States B-1B bomber In September 2017 at the height of the Trump-Kim tensions, Lancer bombers flew north of the demilitarized zone further than they've ever gone before. No response came by N. Korea, suggesting the possibility that bombers have never been detected. Lack of knowledge of various types of weapons that are used in North Korea may make any perturbation to their nation lifestyle, culture, or even honor appear to be an attack. This is how the Soviets took it into consideration in the game of war Proud Prophet.

What are the probabilities of launching a limited strike towards America? United States? If only a small portion of Russia's nuclear arsenal were deployed millions of people could be killed. Explosive and nuclear damage can be dangerous for people close to submarine and bomber sites. In addition, the Atlantic Ocean might be seriously affected when missile silos in the Midwest suffer explosions that break the ground. The crops and animals could be damaged for as long as an entire whole year inside the United States because of the consequences. Even if there was no imminent conflicts and no future conflicts, it is likely that the United States' industrial foundation is most likely to be impacted. In order to cause disruption to an U.S. economy, an adversary could choose to attack one of the nation's most significant sectors instead of pursuing the military targets. 10 Soviet SS-20 missiles with eight 1 megaton warheads, could be anticipated to strike U.S. oil refineries if the attack were to occur. This led to the loss of nearly two-thirds the country's refining capacity for oil. 5 million Americans could have been killed when major cities were evacuated under the scenario that resulted in the attack. The various "limited" nuclear war scenarios would result in the deaths thousands of American citizens, which is a lot greater than 1.2 million killed during all of the battles of our nation. The question lies in whether or whether the United States should are willing to contemplate possible scenarios of slender nuclear conflict. Are we able to be confident that a nuclear war could only kill a few millions people? Strategic planners at the helm are preparing our nuclear response to an enemy's threats. Can we trust them? In the event of an attack is it reasonable to be in a state in nuclear readiness?

What would happen if All Nuclear States broke into War? So long as nuclear states possess hundreds to thousand of weapons that are nuclear, a full-scale nuclear war is a real possibility. Nuclear wars that are all-out can are devastating and have devastating consequences. Cities that are targeted are likely to continue to provide requirements that allow for single explosives. Apart from major cities, most attacks will likely be carried out using one weapon. It doesn't matter if it's an isolated blast as part of larger conflict is not important to those who live near to the site of a blast. The survivors in areas with less damage On the other hand might notice a major change in their lives. Injuries from thermal flashes affect more than the 5-psi blast radius. A single nuclear explosion could result in tens of thousand of severe burns that require medical attention and millions might be injured in a battle that is large-scale. The number of burn victims across the United States number in the thousands, and the majority reside in cities which a nuclear attack could ruin. Burn victims who could have been saved if injuries were due to some isolated event would die in the event of an nuclear conflict. Lacerations and fractures are the two most frequent damage suffered by nuclear blast victims. They also are likely to be affected by other ailments like crushed skulls or punctured lung tissue. A lot of people will die, which could have been saved were modern medical technology readily available. In the event of a major conflict that lasted for a long time, the majority people in the United States would be poisoned. One of the only methods to avoid fatal radiation was to create an escape for those who survived the initial explosion. However many people would be exposed to radiation levels that could weaken their resistance to illness and increase their risk of developing cancer Both of which can be life-threatening. The spread of infection can be to the point of pandemics because of polluted water sources as well as poor sanitation infrastructure. Nuclear war is often referred to as "the final outbreak of pandemic" according to the global organization Physicians for Social Responsibility for reasons that are not related to.

If one is caught in such a devastating situation, it is essential to be prepared to get through. The next part of this article will explain you will find various strategies and methods to keep you alive during a nuclear conflict.

Chapter 6: Be Mentally Ready

In order to survive, you must be aware of the bizarre and terrifying nuclear weapons threats and the strengths and weaknesses of human beings when confronted with combat. The most unexpected threat is nearly always the reason for anxiety, which can be a very unpleasant sensation. In the event that millions of tonnes of earth pulverized were hurled up into the air by explosive explosions from nuclear weapons Some people could believe that an end to the world seemed coming. If they could not understand the images was happening, they might be devastated. Be aware that huge dust clouds, particularly combined with smoke from massive fires, can transform the day into night, like it has been during certain volcanic eruptions as well as the largest forest fires allows people to bear and live. Tremors in the ground and thunder are also common in the unusual clouds. "Artificial auroras" produced by nuclear explosions can be observed in the sky in particular when the explosions occur at a distance of miles above the earth.

Panic and Fear

The fear of death can save someone's life in numerous circumstances. It can increase our capacity to be more productive and work longer when we believe that the end of our lives is near. We are able to do things that which we wouldn't be able to do otherwise since our motivation is based on the fear of death. Inability to perform isn't evident through shaky fingers, wobbly knees or a cold sweat. It's important to keep worries under control by making the work. In the face of the possibility of dying confident and brave people admit their fears. Then, they formulate an approach and begin to identify the root of their fear.

An ordinary person may experience fear and anxiety in the event of a risk that is sudden or if the risk is serious enough. If a person is scared they're incapable of making rational choices. The book, Psychology of Survival, Dr. Walo von Gregor, an experienced doctor with combat experience, describes the two phases through which it develops. If someone is unable to even try to save the family or themselves and their family, they are in the first stage of indifference. In the next stage, you're that you are compelled to go away. Radiation poisoning can trigger nausea as well as vomiting, tremors and diarrhea for people who are anxious or anxious. Terror is described by the term "explosively infective" according to doctor. von Gregor. In the event of an nuclear attack people who have gained an understanding of the human nature of human traits behaviors, behaviours, and signs are less likely be frightened and rendered useless.

The paralysis of emotions

The most frequent reaction to danger is a form of emotional numbness. This may be beneficial. "Emotional paralysis" psychologists say. This is a way for individuals in danger from being overwhelmed by the sympathetic feelings and frightening images. It aids them in thinking clearly and perform better.

The atomic bombs that destroyed Japan during the War to Surrender, were airbursts. As such they did not have local effects. Therefore, we do not exactly how people would react in places that are exposed to explosions on the surface and the fallout. However we can say that it is evident that the Japanese victims' responses appear to be encouraging given the large number of severely burned individuals within them. This is the norm for a population expecting a nuclear attack and looking for any kind of shelter. "Open panic or extremely disturbed behavior only occurred under exceptional circumstances among the hundreds of thousands who survived the two nuclear bomb explosions," says Dr. von Gregor. The same reason applies to the conventional bombs long-term psychological disorders were not common in the aftermath of the nuclear bomb explosions.

In the event of an attack by nuclear weapons Some say America will fall into chaos in which every human being struggles to survive. They do not consider the history of human tragedies that have a lasting impact and the self-sacrificing nature of many people. But America's agricultural areas remain stocked with clean food following a major nuclear strike. Americans living in food-rich regions have always helped those in need. For many Americans to send food trucks to hungry Americans will involve radioactivity as well as other risks. One of the most vital psychological strategies to survive in the current conflict is the belief that many people living in a perfect society will fight for one with one another and cooperate in spite of danger and loss.

2

Alerts and Warnings

M

Any of the tens or thousands of Japanese victims in the first nuclear weapon ever deployed during war were hurled through Hiroshima or Nagasaki were in their large Air raid protections at the time they were hit through by bombs. They were struck off guard by single plane strikes. The fact that shelters are closed doesn't help people unless they're provided with enough time to arrive and be aware of the warning.

Alertness Types

There are both tactical and strategic warnings available.

Strategic alert

The activities of adversaries that are observed as preparations to launch an attack provide the basis of this alert. For instance in the event that Russian soldiers were moving across Western Europe and Soviet officials warned of a nuclear disaster in the event that the countries of resistance began with nuclear weapons that were tactical We would have been given warnings from the strategic side. In the course of a few several days Americans who live in areas where they are at risk of being targeted by terrorists could have had the opportunity to leave. Many millions of us could be able to build or improve our shelters as well as other arrangements in the event of a day or more advance notice. In turn, we could also help to reduce the likelihood of an attack.

Alerting the troops

Our top authorities would be alerted of a nuclear attack against the United States minutes after bombs or other destructive weapons were launched at our country. Our warning systems for military will begin receiving alerts immediately from radar, satellites and other advanced detection techniques. The top-level judgments should be made in evaluating this raw information. The alerts to strike would have been sent out to all communities across all of the United States in the event of an attack by terrorists. Pearl Attacks like those that took place in an area of the United States are far less likely to be recognized by the common American in the event that they precede an alert warning. Although the current warning systems are, it's unlikely that they will be received by the vast majority of Americans enough time to prompt people to seek refuge in the vicinity in case there is an attack. The details of the current alert system are provided in the sections below. Only those who had been well-informed could receive life-saving warnings from the initial nuclear explosions.

Warning by the government (Sirens)

The officially recognized U.S. warning system consists of sirens and radio and television broadcasts designed to send out timely alerts to the public. In order to provide information about assaults to the official warning areas around the entire country the National Warning System (NAWAS) makes use of a wireline network. Nuclear explosions do not protect the NAWAS system from electro-magnetic impulses (EMPs). On designated warning areas authorities will sound sirens in the local area and start broadcasts of television and radio emergencies in the event that power is not lost. A number of local civil defense officials are at NAWAS warning locations. The military's warning and communication systems are constantly being improved as a result of NAWAS is getting information from these systems. The wail-like, screaming sounds that last between three and five minutes are referred to as"the Attack Warning Signal. The signal repeats after a brief interval. There's just one thing to do if you hear this sound and that is to immediately seek out the most suitable shelter available and make sure you take appropriate precautions. Take a listen to the radio when you arrive in a secure location to gather details. In accordance with federal protocols that a foreign attack upon the United States won't cause the Attack Warning Signal to go off. It is recommended that readers review their local plans prior to the event occurs, as local authorities might not adhere to this rule of thumb.

It's crucial to recognize the limitations of warnings about assaults that are sent out by radio stations and sirens:

* There is only few metropolitan Americans who are able to hear the city's sirens , if most city dwellers had fled during a time of crisis.

* In the event of nuclear war it isn't the situation. Anyone who hears the warning signal are not likely to be able to discern it, or to assume that it's a legitimate attack warning.

EMP effects from these high-altitude blasts are mostly designed to disable or disrupt U.S. military communications. These EMP impacts could cause the power grid to be cut off necessary to turn on sirens like the majority of broadcasting stations. This Emergency Broadcast System (EBS) provides public radio warnings and information in the event emergencies (EBS). The system is primarily based on broadcasting AM stations however, certain TV and FM channels are also used as backups. Utilize your normal broadcast frequency in case you're in the middle of a crisis. Warnings about nuclear war will be confirmed by EBS stations that haven't been knocked off by EMP or other explosive effects. People in areas where whistles and sirens aren't able to be heard will receive details. EMP affects on phone systems however will likely limit the amount of data stations have access to. It is likely that the operating EBS radio stations to inform listeners to select the most suitable refuge in their area before ICBMs start to explode. In February of 1986 it was discovered that the Emergency Broadcasting System had only obtained EMP protection for 125 of the approximately 2,771 radio stations within the system. Ten of the existing 3,000 Emergency Operations Centers were also EMP-proof. If there were an attack many radio stations that were shielded could be shut down; the majority don't offer sufficient fallout protection for their broadcasters to keep broadcasting in areas that are exposed to high levels of fallout.

Warnings from the attack

If a nuclear explosion was to occur and the world was struck, the vast majority of Americans would not be injured by the blasts that first occurred. If a major terrorist attack was to take place the initial explosions would allow many people to move to a safe location. The explosions of submarine-launched ballistic missiles will be the most precise way to warn of an imminent attack. Submarine ballistic missiles launched from the submarine (SLBMs). A lot of Intercontinental Ballistic Missiles (ICBMs) are significantly less than that. Strategic Air Command sites and numerous runways at commercial airports are length enough to allow our bombers that are heavy to be struck by SLBM warheads. Naval bases as well as high-ranking commands and communication centers could also be targeted. The majority of Americans aren't aware of how to utilize explosion warnings to safeguard their lives. The only way is to anticipate how an adversary nuclear strike could be planned. Long-range bombers are one of the primary targets for a strike from an enemy as each U.S. aircraft is one of our most deadly weaponry of retaliation. From the air, and away from runways, bombers are extremely difficult to bring down. The decision was made that the first round of SLBMs will be launched concurrently with ICBMs coming from European as well as Asian silos in order to make sure our aircrafts were destroyed instantly. Within minutes of the launch, L.S. surveillance systems would issue alerts. In 1977 there were 210 B-52 runways throughout the United States with this minimum length. Many of these long runways were used to spread heavy bombers during a situation. To eliminate the maximum number of bombers and to deter enemies SLBMs will likely hit and destroy these runways.

Nowadays, the majority of Soviet SLBMs are fitted with warheads of sizes between 100 kilogramons and one megaton. The likelihood is that airbursts could cause destruction to U.S. aircraft and airport infrastructure within 10 to 15 minutes from the beginning of an attack. Even if you were forty kilometers away from the attack, you might hear the sound of glass shattering. Due to this, the majority of people are not harmed by the SLBM attacks. If the general public is informed, these explosions could be used in the form of "take shelter" warnings, which may save the lives of millions.

EMP (EMP) consequences from nuclear explosions may be used as warnings of assault across vast regions, which could cause interruptions in power and communications. The radiofrequency radiation emitted by nuclear explosions triggers an EMP. A variety of electronic devices could end up being damaged, or even destroyed, if exposed to the strong and ever-increasing surges of electrical current caused from E.M. P on long transmission lines. Computers that aren't protected are also at risk of getting broken or damaged. Certain motors and airplanes used in modern vehicles truck, tractors and other vehicles may be destroyed when their solid-state electrical parts fail. Metal is more durable than plastic, however it is less durable. The methods used to protect electrical equipment from lightning-related surges of current are not always effective to protect against E.M. P. Although the precautions are in place, only a tiny percentage of facilities in the civilian sector are shielded from the effects from an electromagnetic pulse. The majority of public power, radio and T.V. transmitting stations that do not have EMP shielding or radios with long antennae could have been affected by three to four nuclear blasts that were detonated at high altitudes across The United States.

Response to attacks that are unexpected

While the probability of a Pearl Harbor-style attack is very low, citizens should be ready in the case in the event of an emergency. Here are a few warnings:

Brighter lights than anything seen before. The majority of Americans could observe the blinding, brilliant illuminations of the initial SLBM explosions that hit targets across diverse regions across the United States. There is the possibility of harming one's eyesight trying to find the source of heat and light even in good weather, from the massive explosions at the distance of as much as 100 miles. A thermal pulse's intense intensity and light will blind you if you do not quickly move away from anything to keep them out. Depending on the length of the flash it could be more than 2 seconds, or as long as 44 seconds when it's from a 20 megaton burst. Get to the door if you're home and you're greeted by the bright lighting. Take a dark path or even the basement if you need to. If there is a shelter nearby but is separate from your house, wait about two minutes after you first notice the light before rushing out to find it. If you're in the outdoors and the light's intensity bothers you, take the most effective protection you are able to. As a matter of precaution to remain in the shade for at minimum two minutes in the case of an explosion's intensity or appearance. The blast waves travel at a speed that is significantly faster than the sound speed when it hits the surface (about 1-mile in five seconds). When its pressure decreases to 1 pound each sq inch (psi) the blast wave as well as its thundering sound have been reduced to a typical speed of 2C/C. It is much higher than the typical rate of sound to this time. To protect yourself from the blast, unless shatters of glass from windows penetrated the skin, you'd require 2 minutes prior to you hear any sound that will alert you that the explosion occurred at minimum twenty miles from you. You can safely remove the most secure cover inside your home for the radio after just two minutes of using it. Find out what's happening by turning the dial towards the station that you typically tune into. While you're waiting, you can make quick plans to move your family to the most secure location you can locate within the 15 minutes it will be taking for the first ICBM to be fired. Do not look out the window or be in close proximity to one once an attack has started. A blast from a hundred miles away can cause glass to shatter under certain conditions in the air.

* The sound of explosions and gunshots. It is anticipated to be that deafening booms from the initial SLBM explosions could be heard throughout the United States. The first time that someone who is hundred miles away from a nuclear explosion is able to hear it will be around 7 hours after the explosion takes place. The majority of people will be in a position to escape before ICBMs were destroyed in the majority of cases.

The power and communications are turned off. A radio powered by batteries can be used to keep dialing even when the lights go out , and several television and radio channels stop transmitting.

If Attack Gets Worst

The effectiveness of warnings could be enhanced if an attack was triggered by a growing crisis. A significant number of Americans could have fled had they been aware that the urban population of the enemy fled or that nuclear bombs used for tactical purposes were being deployed abroad even if our government did not have ordered the evacuation of areas that were at risk. Many of the smartest people would spend the majority of their time in or near to the most luxurious accessible shelters while they slept. While some could have constructed or upgraded small-group or family-style shelters while others would have given them basic needs. This is why the official warning systems should have been fully alerted and improved. Anyone who is in a tight proximity should have an emergency radio available all hours of the night and day during this time. You could inform your neighbors about an emergency situation or an alert for attack broadcast. There are a few disadvantages of waiting until an emergency situation to build emergency shelters. In particular that many of the construction workers will be out in the moment the first missiles go off. Civilian warning systems might not be able at the right time to SLBM warheads' appearance. Wearing caps, shirts and gloves while working outside in the case of an attack can aid in preventing burns. When they first receive the first warning signal or warning, such as the sudden cessation of some radio signals, they must take cover immediately.

You must keep yourself isolated by Shelter

The bunkers may be evacuated for a short period of time following an attack warning, in the event of no obvious explosion or fallout. When most missiles have been fired, bombers from hostile countries may be on the way within just a few hours. Cities, as well as other targets could be destroyed with nuclear bombs within days of the initial attack due to malfunctions of the missile or mishaps. Although the gauge indicates no danger of fallout, the vast majority of people should remain in shelters for a couple of days. Certain people should be exempt from shelters, for example, those who have to remodel or relocate. They could move without risk between radioactive explosions, and then the arrival of aircraft from enemy or the beginning deposition of fallout just after a few hours. In just 12 hours following an enormous explosion, fallout could reach a significant portion across the US. Many people were unable to believe the information from radio stations far away about the changing dangers of fallout and the time and date to get out of their bunkers. Weather conditions and distance, like wind speed influence the likelihood of falling debris. Fallout meters that they own as well as those of their neighbors as well as local personnel from civil defense who've gathered radiation measurements can be relied upon.

Keep your Radios Operating

It is better to have an audio system that is capable of receiving emergency broadcasts. After an attack, the majority of survivors will be distant from radio stations that could be broadcasting to receive reliable information about local, continuously changing fallout dangers. Shelters that have radios broadcasting information on the massive disaster, food relief strategies as well as survival strategies for the everyday and what government agencies and organizations were doing to help improve morale and the likelihood of survival for the long term. It is possible to keep in touch with the world outside after an attack if you can remember:

* To be safe Take your entire family's battery-powered devices and radios to your shelter.

* Only use the AM radio's short loop antennas in order to protect the radio. A dangerous current surge can't be created in small portable radios since the antennas of these radios are too small.

The FM, CB as well as amateur radios must be kept in a short distance, usually less than 10 inches. Don't attach an antenna made of wire or a shorter radio antenna to pipes in the event of an EMP attack that could take weeks to heal due to frequent high-altitude explosions. A surge of current created by an electromagnetic pulse (EMP) is extremely harmful to radios because it can cause diodes and transistors to malfunction , or even reduce the range of their reception.

* If you own an uninsulated radio ensure that it is in a minimum distance of six feet from long

metal objects such as pipes, ducts or cables. Radios that are near conductors made of long metal can be damaged, or destroyed by huge EMP surges that are picked up and carried by conductors.

* So long as it is located within 6 feet of a conductor that is long through which high currents are produced are generated by an EMP surge, protect the device from EMP by completely covering the device in conducting metal. Place the device in the cake box made of metal or storage container, or completely wrapping the device in aluminum foil will shield the device from electro-magnetic pulses (EMPs).

The antenna should be grounded by connecting it to the vehicle's chassis with a wire , when it's not using it. Remove the antenna line from your car radio via the receiver. Metal-to metal contact can be secured by using clothespins or tape. The ideal place to set your center is the frame made of metal on the mirror outside. You should park your car near your bunker as is possible, to allow your car radio to receive distant stations that are broadcasting after the fallout has gone away.

The idea of securing a radio in a clear plastic bag that is large enough to allow it to function inside is a good way to shield it from excess moisturethat can arise because of the continuous use of some underground bunkers.

* For a few months after an attack, you might not be able get new batteries. Make sure to only listen to stations that offer the most current information frequently. The batteries in transistor radios could last for up to two years when played at the lower levels.

3

Evacuation Plan

M

Ulti-megaton Soviet nuclear weapons, including the ones carried by over 240 SS-18s were the most dangerous during the 1970s midway through. Up to fifty percent of these huge Russian warheads were within close proximity to their targets to demolish the missile inside its tightly secured silo. Modern technology has allowed the most modern Russian warheads to strike their targets within a couple of hundred feet. To kill extremely difficult targets, like missiles that are stored in silos that are blast-proofed, this degree of accuracy doesn't require multimegaton warheads. When Soviet defense targets fall, Russian losses in counterattacks are minimized. The military policy manuals written by Soviet Union Marshal V. D. Sokolowski For instance, they emphasize this common sense, long-standing Soviet practice. By following this methodological approach, it was decided that the Soviet Union was obliged to abandon its massive warheads for lesser ones which proved to be more precise. Most large missiles featured several Multiple Independently-targeted re-entry Vehicles by the year of the 1990s. More than ten thousand kilograms of bombs were carried by the majority of SS-18s, along with the majority of U.S. command and control infrastructure, airport runways of more than 6000 feet, and vital ports, in addition to the refineries and industries in the core of our military might be damaged through Soviet warheads. A majority of American missiles stored in silos can be fired upon notice and then be launched into space in a direction towards Russian targets, prior to when Soviet warheads be able to reach their silos, preventing attacks with first strike in some. If there is an ever-growing situation, how will the dramatic changes in the Soviet nuclear arsenal affect your decision to leave or remain? If the warhead was the Soviet nuclear warhead 10 megatons of nuclear weapons that is exploded in the atmosphere could cause a massive destruction of American homes that are up to 15 miles away from the central point. If it explodes as an airburst the result of an airburst from a Soviet bomb believed to be one megaton in size will destroy the majority of houses within a radius of 4 miles. That means that if your home lies within eight miles the possible location, you don't have to move out of the area to avoid dangers from fire or nuclear when you consider the Soviets choice to outfit their nuclear missiles of the largest size with a number of smaller, highly accurate warheads, which is beneficial to the Soviets.

High Risk Zones

The fallout deposition in areas at risk could produce an average exposure of 15000 R and more those who spent the initial two weeks outdoors following the fallout deposition. The most dangerous areas are located within only a few miles from one other and up to 100 miles to the south of the five Minuteman missile fields. According to official risk-area maps of civil defense 120 surface explosions of 25 megatons hit industrial and urban locations were made based upon an unproven assumption. Utilizing only missiles that are air-to-air doesn't make sense from a militaristic perspective, because launching them from the air would cause the destruction of twice as many military installations as well as other industrial and urban assets. The majority of these areas have excellent fallout shelters therefore staying for weeks at a time is an alternative, as is shifting to a place that is typically free of potential fallout dangers. The phrase "very extraordinary" refers to a shelter's capacity to limit the radiation dose to less than one-tenth of what it would be if it been outside during the same time. If a shelter has an effective protection factor of 100 an exposure that is 10,000 R from outside would not be enough to cause a person to become unconscious. However far it's below level, any basement in a home can't provide adequate security when it isn't subject to significant renovations. From 5,000 to 10,000 R for a dose of two weeks outside of areas with high risk of fallout. These areas will require the construction of shelters that are effective in the event of nuclear radiation. Keep enough food and water available for at the very 2 weeks at a minimum. So, those who survived will be forced to stay in shelters for a few more weeks. The risk of radiation in the shaded zones of the map reduces as explosions occur further away. This is typically the case, however snow or rain can transport radioactive particles to Earth.

Further out, there could There could be "rainouts" of heavy fallout. According to the Oak Ridge National Lab's computer-drawn image is not a realistic representation of the dangerous 29 fallout that has been resulting from the surface explosions that are isolated. Although high-altitude winds typically are able to blow from east to west simple fallout patterns like those shown are not recommended as rough guidelines to improve the chances of avoiding an explosive area or a highly dense fallout areas and moving to a less dangerous place. It is difficult to determine what direction the wind will be blowing or where an attacker might shoot and the effectiveness of a weapon be. If you are in an region where radioactive fallout is of concern, then you must take every step to construct a radiation-proof bunker. Incorrectly reading risk-area maps and believing that they're safe from dangers of fire, explosion and even fatal fallout can be easily carried out by people who have not been taught the proper hazards. It is important to remember that risk-area maps that were released in 1986 were based upon impossible scenarios 10 years earlier. Soviet warheads have dwindled in size since 1986, however their megatonnage as well as their fallout capacity remains the same throughout that time. The obsolete attack scenario that created 3,190 megaton warheads for bases for military purposes could possibly explode in our five Minuteman missile fields with almost 2 million megatons. A missile strike against our bases would produce roughly the same outcomes. In the last 10 years, the radiation emanating from the nuclear arsenal of America has been spread out over shorter distances than multi-megaton warheads which would have exploded 10 years ago. Soviet nuclear ballistic missiles generally included warheads of 20 megatons due to the fact that the nuclear fields we have aren't capable of releasing large quantities in radioactive fallout. Even if the current Kiloton-range Soviet warheads were to pour out or snow in the future, it is unlikely that huge quantities of radioactive particles could reach Earth within hours after the airburst and, therefore, the majority of industrial or urban targets would be sprayed by air explosions that result in very little, if any, or incapacity-causing fallout. If the Soviets began a war against America, United States, most of us would be susceptible to fire and explosions than radiation fallout, since we are away from "hard" targets such as missile silos, where significant radiation is predicted to fallout.

WHAT TIMES DO I EVACUATE?

The use of nuclear weapons in tactical warfare were first used in a regular overseas war with Russia. United States and Russia, or Russian cities were evacuated. The removal of cities that are at risk and high-risk areas in an intensifying crisis could increase the likelihood of the majority of Americans being able to survive. It is becoming increasingly difficult for the United States' ability to evacuate citizens in a crisis is diminishing. About 1/3 of Americans living in high-risk areas have evacuation programs in place, out of the more than 2000 required across the US. The evacuation plans of certain cities and states were put on hold by 1986. There remain unanswered questions concerning who would be the person to command any evacuation in case of attack.

Additionally the fact that there is no formal American evacuation plan stipulated the requirement for evacuation in the event of war, and there isn't. Because there is a chance that the United States might be attacked with nukes, if would like to increase your chances of survival, you must act ahead of a potentially critical situation that could arise to improve your chances of survival locally, or prepare and plan to leave. If a war is brewing or emergency evacuation that are triggered by Americans could decide on their own on whether or not they want to go will likely happen. If a mass evacuation is not planned properly, it is more likely to result in traffic congestion and other problems as opposed to a gradual departure of residents from high-risk places initiated by each person for a period of time ranging from hours to days. The majority of Americans living within 15 miles of the nuclear strike's most likely area of attack, the best way to improve their chances for survival would be to stay in their home or close to it and to build or extend adequate shelters in the area, instead of having to move away. These missile ranges, in which many warheads could be detonated close to the surface, and generate significant fallout that could last as far as 150 miles constitute an exception to this rule due to the importance the missile field for the Soviets. Local civil defense organization should be followed, at the very least in terms of the routes and distances to areas that are not likely to be impacted by strong fallout.

Checklist of items for evacuation

In spite of how long the pilot has been flying for in the air, he's still a great pilot. He prepares his plane to take off by running an itinerary. Similar to that anyone looking to leave in emergency circumstances must use an outline of steps to ensure that they have essential things with them. The next section, there is an Evacuation Checklist will assist a family to ensure that vital survival equipment is not left out when making use of a convenient home or basement.

The maps, small radios powered by batteries as well as spare batteries, fallout gauge and writing materials are all part of the survival info package. Pick, shovel and see (a bowsaw is a great tool) and any other tools mentioned in the instructions for construction of the shelter that will be constructed. Wear gloves too.

Based on type of shelter, you'll require waterproof materials (such as shower curtains, plastic, and various other materials) to build it. Unless it's extremely cold and you're also in need of the KAP or other components for your DIY shelter ventilator.

* Measure one teaspoon then add the amount to every cup to check what's right. Also, you'll need all the large polyethylene trash bags that are available along with smaller ones and pillows covers. It's possible that nuclear peace, instead of the threat of nuclear war, will remain in place after the people have left.

There are two clear glass jars that are about 1 pint in size cooking oil; cotton thread; kitchen match and an airtight jar to keep matches dry.

Overshoes, winter boots and heavy clothing (even in summer months, since they'd not be available following an attack) raincoats, raincoats, and shawls are the most important items. Make sure you're properly dressed for the task you're assigned by wearing the appropriate attire and footwear.

* You'll need an extra sleeping bag, or two blankets per person to be able to rest comfortably.

The powder of milk, the cooking oil and sugar are the three most essential baby food options. Foods which are small and don't require cooking are selected. At least 1 pound of salt, a container and bottle opener, a knife as well as two pans for cooking that have lids are required (4-qt size is preferred). A large dish, a cup and spoon must be given to every person in the. It is also possible to build an ad hoc stove from only a bucket of metal with ten coat hangers of wire as well as nails and the screwdriver.

* Toilet paper, soap, baby wipes, tampons and sanitary towel are among the items that you'll need to keep fresh while on an adventure in the wilderness:

* If any member of your family requires special prescription medications, make sure you have the medications available. Keep potassium iodide in your bag to protect yourself from radioactive iodine, as well as additional pair of eyeglasses as well as contact lenses.

Also take two square yards of mosquito nets or bug screen to guard the shelter's entry points if bugs are a problem.

The roads blocked by collisions as well as delays in the movement of vehicles can pose danger for those who have to evacuate. The strength of people can usually clean a roadway when the right direction and guidance is offered. In the time that Japanese were marching towards China I witnessed the Chinese clearing the mountain roads of ruins and other obstacles using the force of their bodies. Monarch Pass that crosses the Continental Divide in Colorado demonstrated that Americans could do similar results if someone could convince them they could achieve it. More than 100 vehicles were stuck on the frozen road after a huge wrecking truck fell over. When I told them how the Chinese solved the same problem but the police officers didn't do anything. The officers contacted the drivers who were stranded to help them get the truck that was flipped back onto its wheels. Around 50 people worked to get access to Monarch Pass in less than 15 minutes. If there is emergencies, the public can help to keep traffic flowing by acting.

AFFORDABLE SHELTER FOR PERMANENT USE VERSUS EV

Instead of running away and fleeing, you can build shelters made of earth and provide them with all the necessities of daily life if you live within or near a large urban region. This is also the situation for those living within 5 miles of a distant area and would prefer to build a secure bunker over leaving. It's a great option for those who (I) the homes are sufficiently far from possible targets and (II) the appropriate space, time and resources are available to build such shelters in a way that is feasible. This Gate Trench may be used as an emergency shelter in the scenario of a nuclear attack. Shelters that are constructed far enough from buildings that are flammable can often be able to save the lives of people victims of fires or explosions. A good fallout shelter that is covered with earth is able to withstand fire and blast damage, but the majority of houses are destroyed by fire and blast prior to that point. A distance of 8 miles of

GZ is predicted to be impacted by a 1MT air blast's impact on fire and blast which can extend up to five miles. The pressure generated by a 1MT air burst could certainly destroy the bunker's footprint of around 100 square miles. This is much bigger than the area of 80 square miles within the four-mile radius surrounding GZ.

Plans for Evacuation of the Pentagon

The residents of in the Washington, D.C., area could be relocated in more rural Virginia or West Virginia in the case that a nuclear attack is launched from the Department of Defense's Defense Civil Preparedness Agency. Six civil defense officers and thousands of other officials from District, Maryland, and Virginia will oversee in the removal of Washington, D.C., along with 400 other nuclear targets throughout the United States. In order to assist in planning evacuations in the event in an emergency the full-time civil defense planner will be hired by Washington's Metropolitan Washington Council of Governments as well as Washington City. Washington. It is estimated that the Pentagon estimates that it will be $50 million. In the current year the Pentagon is planning to spend around $7 million. Residents will be relocated into "host" counties within Virginia as well as West Virginia, some of that were more than 200 miles from each other, during a period of up to 2 weeks stretch. The residents of Loudoun as well as Culpeper counties could suffer from the change. Anyone within a 15-mile radius within a 15-mile radius of the downtown area D.C. is likely to be evacuated in case of nuclear war and transferred to facilities, either private or public, which provide some protection from the radioactive fallout. In towns as small as Dover, Delaware, nine evacuation tests have been conducted within the past three years. In the next few months, the latest version of the evacuation plan will be in place. The evacuation plans are impossible in the eight million-person island city in New York City. In the 1950s, although fewer people had to be evacuated and nuclear bomb-carrying aircrafts were less powerful, D.C.'s evacuation procedures were however designed to be complete and thorough. Because newer and more powerful missiles were being created and tested, there was no plan to put these ideas into action. As per John E. Bex, who lives in an underground bunker made of concrete, which is located under a cow pasture situated in Olney, Maryland, a one or two days' notice is "absolutely feasible" to evacuate Washington, DC. In the peak hours, Bex and his 360,000-strong team of workers are on the job every day. A underground communications facility worth $14 million constructed in 1971 is available to both civilian and military personnel. In the case of a nuclear strike residents of East Coast would be able to evacuate in a matter of minutes. East Coast would have just 15-30 minutes in advance According to Pentagon sources. Therefore an exile from Washington DC would only be needed with "days or two days' notice." Many military experts believe that a nuclear war is only likely to occur in the middle of a global emergency, like in 1962 during the Cuban Missile Crisis. The authorities within the United States would have enough time to evacuate cities if this plan were accepted. It could be a retaliation for having announced an evacuation plan for Soviet cities. A few aren't. It could be the start in the nuclear conflict supposed to be avoided as Aspin claimed. The civil defense system within the United States has been a "duck and cover" operation since the year 1960. People go to fallout shelters in the case of nuclear war. Even if nuclear radiation did not kill you immediately there could be months of radioactive dust in the countryside. For over fifty years civil defense officials have Civil Defense has been striving to increase the number of fallout shelters as well as strengthen communication and emergency warning systems within the Washington region.

The amount of supplies totaled around $150 million in 1962's Cuban missile crisis in 1962. Civil defense in the local discussion revolved around the supply of these items and whether local authorities needed to have permanent civil defense personnel. To cut costs, Arlington City Council has looked at dissolving the city's 145,000-per-year office for civil defense, which has a staff of 14 that the council been debating recently. In order to save $420,000 in salary severance payments the council decided not to shut down the office. In the event of any kind of weather, from snowfall to natural disasters, to large-scale events such as parades and inaugurations D.C.'s Office of Emergency Preparedness (formerly the Office of Civil Defense) is responsible. The office receives greater than $250,000 in match funding from the Pentagon to help with civil defense issues. Civil defense and law enforcement communications centers are situated in the Municipal Building 24 hours a all day, 7 days a week. To aid states and communities in preparing and responding to natural disasters, such as storms and floods in the event of a storm or flood, President Nixon required to create the Pentagon Office of Civil Defense to change its name to"the Defense Civil Preparation Agency in 1972. Following the congressional vote last year and this year's Ford Administration was not able restrict the authority of the agency to nuclear war readiness.

Due to military orders to eliminate emergency food and medical supplies kept in shelters last autumn, the civil defense function were shut down. It was never intended to be consumed by humans, as per researchers. Prince George's County, Maryland farmers received many tons of survival biscuits from 1962 as part of a huge humanitarian effort to provide food. Other options for disposal include recycling or donating the proceeds to hospitals in local areas. Despite their unpleasant taste the biscuits are considered acceptable by some civil defense authorities. Survivability biscuits do not pose a risk to the human diet according to the head of DPR. As per popular opinion they were banned by the Center for Disease Control banned biscuits in the United States after it received an example. In 1955 the president Eisenhower "flew" his capital, Washington D.C. in a convoy of 15,000 federal personnel to conduct an attack with nuclear weapons on the United States.

Chapter 7: Ultimate Want: Shelter In Isolation

The ability to stand up to the threat of nuclear war requires an ongoing defense. The majority of nation might benefit from a sanctuary that is immune to radiations from fallout and to the elements. The majority of buildings are not able to offer adequate protection against fallout if the wind is not in the right direction during the deposition time in most areas that aren't at risk of being hit by a major nuclear strike. People who live in or close to an urban area, or another possible target area, requires greater protection against explosion or fire as well as fallout. The fallout shelters discussed in this book are able to withstand the force of explosions enough to destroy the majority of houses, according to research. If you follow published, field-tested guidelines, tens of millions of Americans can construct these fallout shelters that are covered with earth in just 48 days in a matter of hours or so." A lot more time, money, and knowledge are required to construct high-quality blast shelters than fallout shelters. When it comes time to build life-saving systems and emergency shelters for the average American must know precisely what they need to do and the reasons why it's beneficial to build them. We're not the type of people who simply go along with what the boss tells you. There wasn't time for long explanations of every element's significance in a time that could have led to nuclear war.

Let us look at an example where a family living in a city employed these directions to construct an emergency shelter within 36 hours after having been evacuated. The report that follows explains the ways in which these guidelines could be utilized to safeguard yourself from radiation-related dangers. The family in question was among many who were enticed to build shelters or life-support equipment for a cash reward equal to the amount earned by labor. In the earlier chapter, the evaluating period began when the family received the written and verbal instructions prior to leaving in an car. Other families who participated in the test were reimbursed for all their goods and this was not the only one. Families who put in the effort and completed the work within the allotted time received a financial reward. The workers were primarily guided by written instructions during these tests, which were constantly improved. Many millions of American citizens worked hard to build shelters that were covered with earth in the event of a nuclear war, which will be more secure from explosions, fallouts and fire than but a tiny fraction of the structures in place. Two conditions must be met in order for this to happen First, our nation's top officials must offer an inspiring, empowering leadership during an escalating, severe crisis and, secondly, Americans must have gotten beforehand, well-practiced and tested survival guidelines for building an appropriate shelter.

Protection from radiation

It is essential to prepare to defend themselves in battle. One's NBC training for defense and the use of terrain and shelter are crucial. For soldiers, knowing how topography impacts nuclear bombs could determine the gap between or death. Knowing how to identify defensive positions which protect people from nuclear weapons is possible through the right training and experience. The slopes that revert in the direction of mountains offer some protection from nuclear radiation. The mountains and hills are prone to absorb radiation and heat from the nuclear blast's flame and also from the first radiation. Due to the slope all projectiles that go over the troops will be redirected upwards. The risk of nuclear accidents can be reduced by coverings like ravines and gullies as well as natural depressions, such as holes caused in fallen trees. However, knowing the precise whereabouts of an enemy nuclear strike is a challenge. An attack from a friendly force gives soldiers additional time to prepare. The best defense is underground with roofing over it. Smoke can be used to limit the thermal energy effects of nuclear explosions in an active nuclear atmosphere or in cases where the risk of nuclear weapons is high. If an officer is able to access Smoke generators, they might be able to provide this ability. It is crucial to act prior to an attack to improve your chances of survival. Choosing the right shelter, strengthening it, and protecting essential things are just a few of the steps listed on this list. The survival of a group could be enhanced by these steps prior to preparing. Make the defensive postures for your group whenever you can, in accordance with the scenario of the day. From basic self-defense positions to more sophisticated ones, you'll discover everything in this article. The protection against nuclear impacts can be accomplished in nuclear environments through fight stances as well as enhanced postures.

Protection from radiation should be of top importance especially from gamma and neutron radiation. To shield against gamma radiation massive layers of heavy or dense shielding material should be employed. Iron, lead and stone are only some examples. However, light hydrogen-based materials provide great protection from radiation from neutrons. Paraffin, water, oil, and paraffin are just some of the examples. Neutrons can be absorbed into specific substances, which results in the production of extra Gamma radiation. This radiation must be shielded from. The more layers of the shielding material, the greater the radiation protection. The following paragraphs will discuss the measures to protect against nuclear radiation. The best defense against nuclear radiation is to dig deep. This is because of the earth's excellent shielding properties. The initial nuclear effects are effectively protected by a properly constructed location for combat. Radioactivity residual can be reduced because of this (fallout). When possible, soldiers should reinforce their foxholes/fighting areas to resist the blast waves. Size of opening to the area could be decreased by the lining of foxholes or by riveting them. The radiation that enters through smaller holes is less persistent and initial than radiation coming through the larger ones. However, many metal surfaces can be excellent heat reflectors, which makes them perfect for insulation. Keep these surfaces safe from the heat of nuclear explosions by protecting them with. A smaller foxhole would be more preferable. The bottom entrance of a foxhole is the place where the bulk of the radiation from gamma is absorbed. Because of the smaller entrance, up to two times less radiation can enter the foxholes of one person than foxholes for two people.

The radiation protection is superior in a foxhole that is deep or a battle location than an area that is shallow. It protects the occupants from nuclear radiation due to the increased thickness of shielding material, or earth. As a result, it decreases the amount of radiation that gets into the hole. The foxhole's depth of 16 inches reduces radiation by two-fold when a person is in a combat position. A four-foot-deep fight posture offers up to six times the protection than a more shallow one. The reflection or line-of-sight effect of sidewalls can permit foxholes' troops to get exposed to heat. Utilize dark, rough materials to cover reflective surfaces to protect troops and equipment from injury. Examples include blankets made of canvas and wool (such for half-sized shelters). Be aware that these materials could remain charred or burned in the presence of high temperatures. Avoid direct contact with them. Ponchos as well as other rubber or plastic-based foxhole covers should not be used solely for cover for a foxhole. They may be a source of fire and cause injury. The radiation from thermal sources is cut by about half when a foxhole has been lined with standard metal screening. The screening can offer thermal protection, but without blinding soldiers' vision through ports. Soldiers should remain at a low level and cover any areas of vulnerability. The risk of exposure to radiation from nuclear sources is diminished by lowering the temperature of one's body.

Types of Shelters

Barrier Type

Trenchways that are open would be filled with fallout in the real-world catastrophe zone). The concrete that is 3 feet thick or the earth depicted in the picture will absorb most of the gamma rays from the fallout. One particle of fallout, approximately 2 feet away of the edge of the trench is the only thing that's considered in this situation). If only one gamma-ray out of thousands of particles passed through the 4 feet of soil and hit the man who was in the trench this was an unique incident. Because of the depth at which the earth is in between them and the depth of the trench, rays coming from fallout particles located 4 feet or less away are not significant. Rays from fallout particles that are 3 feet or less away, but are diminished due to the distance. It is impossible to shield trench dwellers against "sky shining," resulting from gamma radiation scattering when hitting oxygen, nitrogen and other atoms within the atmosphere. Since his head is higher than the ground The person will be exposed to about 10-percent radiation. A person who is exposed in an open trench could be exposed to dangerous amounts of sun's rays from nearby heavy fallout regardless of the amount of material that is removed from the trench. The sun's rays hit the planet in every direction. If the sun were to be in a more deep hole, certain radiations would be shielded from his presence however, not all. Barrier shielding is needed to protect you from any direction.

Geometry shielding

If you are using geoshielding, you can increase the distance between you and debris falling on you and then alter the entrance holes of the shelter. Avoiding light sources will minimize any light radiation that could cause harm to your. If you're located in one of the lower levels of building that is tall it will give you less radiation from falling particles that if they were on the floor directly over you. Like gamma radiation, when light travels through a long tunnel through an opening at the far end it will be less able to get to you at the other side than it would in a narrower corridor. Turns in corridors significantly restrict radiation that enters a shelter. Right-angle turns slow down speed by 95%, regardless of how they're started either horizontally or vertically.

Covers for Overheads to be used in the Field

By using an earthen or other material-covered roof, thermal as well as nuclear radiation in the early stages and fallout can be reduced. A roof with an overhead covering can prevent collapse. Missile defense can also be provided. Overhead shielding low-quality should be avoided. To ensure that the shield will stand in such a scenario it must be extremely sturdy.

Wood, metal pickets and various other materials can be used to construct the barrier that is covered with earth and sandbags in order to shield the structure from intruders. The use of earth-filled ammunition boxes for an option for protection is an alternative. When you design an appropriate overhead cover, bear these aspects in your head:

* A material that is heavily coated is the best.

* Make sure that you have a an even base.

* Try to fill as much space as possible.

A truck could quickly and effortlessly provide overhead protection. It's an easy and quick method to escape the trap. The vehicles with armored bodies offer a distinct advantage over those equipped with wheels (of course having the vehicle with armor is more secure). Even when an overhead cover protects the battlefield radiation can still penetrate the battlefield and cause harm (between the treads roads, wheels or tires). In certain locations Sandbags can be utilized to cover the area. For the sake of reminder the car does not have a robust neutralon shielding ability. The displacement of the vehicle and foxhole collapse are other possible results from the explosion wave.

Utilizing Earth as an Shield

If a nuclear blast hits the inhabitants of bunkers and other structures that are protected could be protected from serious injuries or even death. However, there are still concerns about the penetrating power of radiation, however. Radiation is everywhere following an explosion. The closest people to the fireball however are able to fly on straight lines. The soldier needs to be protected from the blast by the maximum amount of soil feasible. The greater the protection the greater amount of soil cover there is. It is possible to get an eightfold boost in protection when you fight in an open position. Radiation scattered through the air can be absorbed however, the majority of direct radiation is blocked. Six inches greater coverage of the ground in the sky reduces scattered radiation two. The use of a flat earthen cover to safeguard underground shelters is much more effective than using the same materials to protect an aboveground structure that is the same thickness. The line of sight in the underground is more dense; this is the reason.

A second set of sandbags guards the battlefield areas. The filling of each sandbag with sand or clay that has been compacted reduces radiation transmission by two times. Sandbags that are thicker, generally will protect you from harm more efficiently. Check that the sandbags are not full of cracks, to avoid leaks of radioactive material. Neutron radiation can be controlled. Gamma radiation that is released by water is deflected and absorbed by water, but a significant shielding is needed. Concrete or wet mud acts as a shield against both types of radiation. If the concrete is 10 inches thick or the damp earth 20 inches serve as shields, the radiation exposure is cut by 20. Ratio of reduction is 2 for every 5 inches of sandbags that are wet. Containers made of gasoline, water or oil can also serve to shield neutrons quickly. All parts of your body need to be protected from radiation exposure. The sun's heat must be kept out of Sandbags. More likely is to find that the contents in the sandbags will be removed due to the blast wave when they melt. A tiny amount of sod or dirt can be added to sandbags in order to stop this from happening. To further conceal and protect against fragmentation that is typical Sandbags could be coated with a variety of substances.

Alpha and beta Particles Protection

Fallout particles not only release dangerous gamma rays but they also release many other harmful innocuous products. The risks of alpha and beta particles are identical. If people are aware of the dangers and have taken refuge in almost any shelter prior to the fallout begins to accumulate in their vicinity it is unlikely that they will be seriously injured by the radiations.

Beta particles, which are electrons which move at high speed are released by certain radioactive elements found in fallout. The fallout layer protects buildings from the outside, rendering it difficult that beta radiation penetrate inside. Food and drinks that are contaminated by fallout and fresh particles from fallout pose the sole threat for humans. Freshly produced fallout has an extremely high amount of radioactivity because it's only several days old. New particles can remain on the skin for longer than a couple of minutes and result in a beta burn. If only the fresh particles of fallout are shielded from skin, the thin layer of clothing could delay the beginning of beta burns for an extended amount of time. Beta burns can occur by fresh fallout particles together with sand and dust get into the nose and ears during the conditions of dry, cold, and windy conditions. Beta burns can be avoided by washing the area promptly. Simple rubbing and brushing can eliminate any fallout particles that remain off the skin, if you do not have access to water. Eyes and skin can be a little harmed by beta radiation that is released by newly fallen particles of fallout on the ground when exposed to a large amount of fresh fallout. Simple eyeglasses and clothes that shield body from radiation can effectively protect the eyes and the skin. Regular clothing can shield and shield the body from any radiation except for the most powerful beta particles released by new clothing. Before entering a shelter, one must scrub and beat their fallout-infested clothes to get rid of as much dust as is possible.

Beta burns in humans were first discovered in the aftermath of an incident that occurred during an incident during the United States' largest H-bomb test in the equatorial Pacific. The meteorologists didn't know how the fallout would finish as it blew in a surprising direction. Within five hours following the megaton breakup of the surface, Marshallese people saw a white powder deposition over all exposed surfaces including their bare and moist bodies. They didn't realize that the small particles were particulate matter that had accumulated of a nuclear test. A portion of it is made out of sand. However, the majority of it is limestone and coral reef powder. debris.) The indigenous people have mistaken the white dust as volcanic ash by the indigenous people because of their ignorance of the fallout. Before they were taken away from their homes on the island and treated by a doctor, it was not visible to the natives. Following two days exposure they were exposed to an estimated dose of gamma-rays of 160 R due to the fallout.

Children enjoyed playing on sandy beaches that were contaminated through radioactive fallout. After a few days the fallout itched and burned the scalps of islanders' feet, and necks. A few days later beta burns and a lot of skin discoloration started. The healing process for burns caused by beta is long even though they're not life-threatening. Several individuals developed ulcers despite getting great medical treatment. But, there was no lasting damage to the skin due to. The inhabitants of sanitary, tight and polluted shelters might be more susceptible to beta burns than Marshall Islanders were from the radiation. After a while, they started experiencing thyroid problems because of the accumulation of radioactive iodine inside the thyroid glands following the first exposure. (Protecting yourself from the long-term dangers of potassium iodide is easy and cost-effective, offering 100% protection for a low expense.

The radioactive atoms that make up fallout possess nuclei that look like the helium atoms leading to the formation Alpha particles. They aren't able to penetrate until there are three to 4 inches air space between them. However thin the substance might be the alpha particles are unable to be able to penetrate it. Only when they enter the body and become stuck in tissues of the lung or any other organs, can be a danger. Any shelter that can fully rid of fallout particles is in a position to greatly reduce risk of radiation. If the survivors do not consume or drink water or food that has been exposed to radiation and alpha particle poisoning is low.

Basement Shelters

In an all-out attack that was the size of what happened in the 1980s American structures and homes were destroyed or damaged, placing the residents who live within the structures at danger. Shelters used within structures that were not within the explosion or fire zones would not be as risky. The Soviet policy doesn't appear to target metropolitan American exiles living in areas prior to the invasion however, an adversary could be able to do so. Shelters made of earth within a blast zone cannot be equaled by basements for security. If an nuclear war were to occur in the near future, it's highly unlikely that the majority of urban residents across the United States would be able to build fantastic quick shelters that are separate from structures and surrounded with earth. Therefore, in the absence of better shelter for urban dwellers unprepared will be forced to seek refuge in basements or other buildings that are already in use.

The emergency shelter of a building like basement emergency shelters, should meet the same requirements as a normal emergency shelter, including being adequately protected from radiation from nuclear fallouts, adequate ventilation water and cooling, and other basic necessities like food and sanitation. Publications on civil defense, like "In the event of an emergency" or "Shelter in the Nuclear war," provide brief sketches and descriptions of ways to improve the fallout safety that basements of homes provide. In 1982 the millions of leaflets were available to be distributed in the event of a national emergency.' Due to the fact that the megatonnage as well as the quantity of Soviet warheads was so smaller when these guidelines were formulated and are no longer in use. In accordance with official civil defense guidance that is currently accessible to the majority of Americans and the different types of shelters that you can build yourself do not specify what degree that radiation shielding they offer (what level of protection they offer). It's not clear how to provide the proper cooling system vital for families that live in a basement during summertime. Food, water and shelter improvements within the home and other basic necessities to live are all distributed with outdated or inadequate details. Imagine if even a small impact shook the house due to an explosion that was far away. The strengthening may be needed to ensure adequate overhead shielding in areas that have a lot of fallout. In this article, we will discuss how a typical basement could be made more effective in mitigating the impact of nuclear fallout in the following sections. A basement could provide excellent fallout protection to a number of homes if modified in this manner.

Then the foot or two of dirt must be spread on the surface to cover the excavation area. Pillowcases or sacks can be used to efficiently move the dirt. Because of the lack of digging equipment, heavy items are best placed on top of the ground in the event that the ground has become frozen or otherwise inaccessible. These materials should be heavy enough to provide a 90-pound-per-square-foot load, which is about the weight of a foot-thick layer of earth. In normal circumstances, the floor joists will be supported by posts, which would be erected later. In the final phase the floor joists are supported by an horizontal beam. Then, you can add an additional one-foot-thick layer soil over the previous layer. To guard the basement from radiation from fallout all windows, excluding one, should be covered with a board as well as all aboveground portions of the basement walls should be covered by 2 feet of soil.

A hand-made air pump must be used to provide adequate airflow and cool. The process of improving a basement takes more effort and money than building a simple shelter in a trench that is covered. If you're seeking the most secure protection from radiation, explosions, and fire, an earth-covered structure separated from the building is the ideal alternative. A small, secure shelter could be constructed in the basement if the family doesn't have the resources to build an earth-covered, separate shelter outside. These kinds of indoor shelters are not more than three feet , and no more than 450 inches tall for those who are tall. Furniture like benches, chairs as well as crates and drawers are a great option to construct walls. A strong roof can be constructed with inside doors. You can also make quick water containers from garbage bags made of plastic that are heavy duty or 3 mil polyethylene film, and cover them with it. For this, you need to place plastic wrap inner surfaces of desk drawers, containers garbage cans, pillowcases and other similar items. Make your shelter by putting in the containers that line them and adding water to these containers. drinking water.

Ventilation of Shelters

For many days in warm or hot weather, occupants of high-protection-factor shelters or most other shelters that lack appropriate forced ventilation would be in danger of heat exhaustion. If they were forced to stay inside in the summer, it could be too hot and humid to live. In a cramped, long-occupied shelter, the body's heat is released by the heat as well as water vapour which could be fatal in the event that fallout stops people from leaving the shelter during summer when people are forced to go to the basement or refuge. The air at first is pleasant and nippy. After a couple of days, even a shelter that is cold could have absorbed all its body heat that it could retain because of the absence of adequate air circulation. Certain shelters can reach dangerous temperatures in just a few hours, unless the air that flows through the shelter is able to remove majority of the people's bodies heat as well as sweat evaporates. Hot or warm weather can increase the risk of high humidity and extreme heat. The ability to keep shelters that are inhabited well ventilated and cold enough to last for extended periods of hot or warm conditions is among the most crucial capabilities of survival in the event of nuclear war that everyone should learn. Ventilation systems that are homemade and methods to stop falling out particles of air from entering shelters comprise a few options. Guidelines for directing airflow into shelters by the most simple method that is possible that is directionally fanning

Utilizing a huge-volume shelter-ventilating pump is the most efficient method of making your shelter in good condition during hot temperatures. This pump was designed in the year 1962 by its creator. In homage to the traditional "punkah" fan that some primitive people in hot climates still make use of, I named this pump the punkah-Pump. The air pump, unlike the punkah can move air, in contrast to the punkah, which isn't able to. This air pump, unlike the punkah Kearney Air Pump (KAP) was tested for the first time at The California

Institute Of Technology, the Defensive Systems Development Center, as well as The North American Freight Company. It was designed to spread air inside shelters and circulating those inside The pump was proven to be efficient in these tests.

Shelters for hot Weather

With the humidity and heat of the outdoor air of the day, this Navy test revealed how much current Americans who are accustomed to central air conditioning can learn from indigenous people of the jungle in terms of being cool and healthy. Natives' methods of avoiding unhealthy conditions like those that were found inside the Navy bunker and the way it was vented in the typical American manner, was evident to me. In the same way I was working on the jungles of South America. There are six methods to stay cool in jungle and these are just the first five. In the course of 24 years of research into civil defense we've developed and tested a pump made from wood that significantly improves the cooling capabilities of people in the jungle.

This pump is able to maintain an enduring temperature even during the hottest part in the United States, assuming the tests have shown that the parameters listed below are all met in a completely populated underground shelter. For the majority of underground shelters as well as some above-ground shelters cooling methods outlined in this article can be employed to maintain livable temperatures during hot conditions.

A greater than 2-degree Fahrenheit increase in the exhaust's "effective temperature" is not recommended near the exhaust of the shelter. If the air is at the absolute humidity that is 99 percent or more, it's considered to be at an "effective temperatures," that is, the point at which an individual experiences the same amount of cold or warmth in their body. The "effective temperature" is affected by variables like humidity, temperature and motion. The"effective temperature" cannot be measured with a traditional thermometer. In the course of testing for occupancy, when the outside air supply was dry and hot those living in shelters that were crowded found that they felt warmer in the air exhaust end of the shelter than their air intake end. Despite the fact sweating had reduced the ambient temperature, the higher effective temperature was a certain indicator of heat exhaustion. The temperature of the shelter air should not rise greater than three degrees F when the temperature of the air outside is typical of the most scorching hours of a heatwave an area that is hot and humid located in the United States. A majority of people are able to tolerate temperatures that are just 3 degrees Fahrenheit warmer than the temperature of the surrounding air in the event that they are sick and cannot function without air cooling.

Temperature of air must not be increased simply by moving it. In an Navy test, the frictional resistance of the filters and pipes together with the high-speed electric ventilators have increased the temperature that is delivered into the shelter 4°F. In the event that less than cubic centimeters per person is provided and body heat increases the temperature of the air by several degrees, a four-degree temperature difference between the outside and inside air can be catastrophic in extreme temperatures.

Check that the airflow of the shelter is uniform. Air is pumped from one side and out from a trench-style shelter in order to make sure that airflow is equally distributed. A huge shelter, like one in a basement, could be is ventilated by a single source as well as exhaust air vent. There is no cooling available to anyone who is not part of the airflow.

Salt and drinking water should be available to residents. If it's hot out, you'll have to take 5 liters of fluid each day and add a teaspoon of salt.

Natural Ventilation

If large apertures are provided at both ends of an elevated shelter in the event of a breeze that is strong enough, air will be forced through. Conditions of temperature and humidity could rapidly change in the event that the shelter is full with water and the temperature is warm.

Natural ventilation can be more difficult to achieve in shelters that are subterranean. A trench or any other shelter that is partially obscured by supply and people isn't a common place for air to turn right angles and drop down a vertical shaft and then complete a right-angle turn. When people are inside shelters during cold weather and the air inside the shelter is more drier than the outside air. It is best to have an air vent that resembles a chimney on the ceiling and an air intake vent is opened near the floor, allowing air to exit the shelter. There is a chance that, in winter, natural chimney ventilation could be enough to stop carbon dioxide levels from being dangerously high. A good example is the igloo that was constructed in the Arctic by Inuit people from the Arctic. In the summer many long-term tenants of shelters with high-protection factors are able to see that natural ventilation of the chimney type isn't sufficient. If up to 40 cubic feet per person is needed in the body-warming shelter isn't any more or less than the outside air. Chimney ventilation is ineffective under these conditions.

Chapter 8: The Food And Water

Nuclear explosions can cause radiation contamination in food items and water. Due to the public's reaction to nuclear tests, the civilian and military populations will be faced with water and food pollution problems. But the military will take on these duties not only for military personnel, but as well for civilians. The military will make supplies available to civilians if the food supply of civil society has been destroyed in specific situations. In this regard, it is crucial to evaluate the situation holistically, taking into account the full scope of the issue as well as the capabilities of civilians and military assistance available. In addition the military should be prepared for the issues that could be faced if nuclear weapons were employed to their fullest potential. They should prepare for the possibility of an indefinite nuclear war. In the end it is possible to estimate about 10,000 megatons worth of nuclear energies will go off within the course of a year. The risk of long-term exposure to strontium 90 should also be taken into consideration.

Strontium is similar to calcium and is a well-known component of our diet, and follows the same pattern.

Water Requirements

A few of us within America United States have ever had to endure extreme thirst. It's an accepted fact that we'll be able to keep our water filled. We often consider "food as well as the water" as we imagine the necessary items. If there was a major disaster people who aren't prepared would realize that conserving water would have been their primary concern.

The average person has to drink at minimum one pint of fluids each daily to allow their kidneys to efficiently eliminate waste. The majority of people drink enough water to flush two quarts if there is no shortage of water. The loss of water is caused by sweat, evaporation, or excretion.) 3 quarts of water per day can sustain an individual for weeks in frigid temperatures if they consume only tiny amounts of food, and is low in protein. However, cold temperatures could be an exception in areas of underground shelters with a high density which were used for a long time. In extreme heat the equivalent of four to five quarts of water per day is required without a provision to wash. It is recommended to have enough water available for everyone who is in shelters for two weeks , to drink. In general, this amount of water will last two weeks, which would prevent drinking if there were any fallout dangers.

In the two-week Navy housing occupancy study in the year 1962 99 sailors each consumed 2.5 Quarts of drinking water per day. The test was conducted during August at Washington, D.C., with unusually cold conditions for the season. The only exception was on the last 2 days of the experiment did the shelter get air conditioning. Salt-related symptoms, including cramping, can appear within a couple of days if someone sweats too much and is not eating food that is salty. A daily intake of 6-8 grams (1/2 tablespoon) is suggested to prevent this. Even if a person eats only a small amount of food that amount of salt must be added to their water. In the heat salt can enhance the taste in the drinking water.

How do you fetch water with you?

The majority of homes have only the capacity of a couple of large containers that are able to carry water to shelters and then store it for some weeks. Alongside garbage bags waterproof lines for cloth bags and pillows are also available. Beware of chemically treated insecticides or odor control plastic bags. To close a plastic water bottle first, spread the top of the bag so that the sides that are inside meet. Fold the opening four times to smooth the surface. Continue folding the opening across the middle to create five inches of wide opening. In the end, fold your bag's bottom upwards to ensure that the opening is closed. Bind and secure the folded-over portion with a bow by using fabric or a lighter cord.

Be sure that the plastic bags that hold water's apertures are greater than the levels of water inside. To transport this type of water bag inside the vehicle, you must put a rope on the bag's opening, and then secure those plastic baggies precisely below the holes that they have tied shut. On the other side of this rope needs to be secured to an object so that the openings remain above the water. The larger plastic bags could contain more liquid when 2 pillowcases or bags are linked. Before tying them together, place pebbles or other small objects in the perforations. The pebbles should be tied at least 3 inches lower from the opening of the bag in front than the rear bag that will be carried. This will prevent the burden from putting the stones on the shoulders of the bearer. If you are unable to find pillowscases or other fabrics or bags, a pair of jeans that have both legs tied tight could be a good way to distribute the burden. A weight that is balanced and could be carried over the shoulder instead of the back eases lower back as well as the shoulders. In contrast trouser legs are just too thin to carry more than the size of a few liters. Keep the levels at a minimum to prevent water from flowing through the holes that are tied shut.

Storing Water

The level of water within bags must be kept at a level below the apertures when making sure they are kept in the shelter. A stay of two weeks in a decent shelter will require a large storage area for water. Pits that are coated with plastic and excavated close to the shelter can be a safe method of storing huge quantities of water at a minimal cost. There's nothing better than this to shield the pit bags the top edges from being sucked into. Create a wire hoop of the identical size to the bag's opening, and then tape on the outside of bag. If you have a strong ground, you can secure the sides that are turned under of twin bags using six nails long, driven into the ground.

Watertight "hidden roof" protects rainwater from being contaminated by stored water, which could contain radioactive substances or germs due to fallout. The thick earth covering above the flexible roof protects against blasts by bending in the event of blast pressure. It was the case that Defense Nuclear Agency tested an overpressure range that could destroy even the strongest aboveground structures with a complete water storage hole similar to the one in this.

A more efficient method of storage of water can be found here. The bag's entrance can be closed to prevent leaks if the soil is unstable enough to create a non-sharp water storage pit with vertical sides. The bag should be filled with water, then tie it up. Then , fill it up with earth up to the level of the water. The loss of soil's pressure along the seams may result in leaks. It is important to note that the weight of soil that is on the roof will not cause the bag to break and leave an air space. When the ceiling and earth are removed, loose dirt is pushed into the bag and compresses it over a lower level of water which makes it difficult to remove. This method of storage is not without flaws:

Many water bodies can be stored in rectangular holes. Pits must not be larger than 3 feet, allowing the possibility of covering them with wooden poles or plywood. The pit was lined with a 10-foot 4 mil polyethylene sheet. By making tiny holes into the plastic sheet and filling them up with dirt helped ensure it stayed in its place. It was laid 40 inches deep, with an polypropylene "secret roof." Earth as well as it's "hidden roofing" were similar the pit cover. About 200 gallons worth of water were kept in this trench. When the pit was subjected to blast impacts sufficient to smash massive structures the water was not leakage. Because of sidewall collapse which led to the leaks. Pits that were rectangular that had higher pressures failed.

In contrast to most plastic trash cans, the metal cans aren't waterproof. Watertight trash containers and wastebaskets are able to be used for emergency water storage, provided they are they are thoroughly cleaned and disinfected with the powerful solution of chlorine and bleach. Making use of fresh plastic film to create an ointment that is waterproof for any solid box is an alternative. Containers made of rough material should be lined using fabric in order to decrease the chance of puncturing. Following an assault the water that was stored in water heaters, toilet flush tanks or tubs might be available as an emergency source of drinking water.

Food

Americans are familiar with eating frequently and in large quantities. That means that he might not think that the first 2-3 weeks after an nuclear attack are not vital for survival for the majority people. There are some cases of exceptions, like infants, young children as well as the elderly and sick people who could end up dying within a week, in the absence of adequate food. A proper protection against explosions and fallout risks, adequate air flow, and a sufficient drinking water are far more crucial than any other aspect for long-term survival.

It's possible that an typical American does not realize that a few easy food items that aren't processed could keep his body going for months, or even for years. Maize and wheat could help maintain a person's health for months, if they are able to cook and eat whole grain grains like maize or wheat. The grain diet can be improved by adding beans to ensure many months of good health.

It's based upon ancient methods to prepare and cook the staple beans and grains which most modern Americans don't know about. The Oak Ridge National Laboratory's Civil Defense experts have improved and tested the field techniques. In the event of a massive nuclear attack was to occur and there is a chance that United States' food supply network would be completely destroyed. The majority of our high-protein diet could be destroyed in the event of a massive catastrophe and catastrophic fallout. Farmers were unable to feed their livestock. If the owners do not take care of their animals, livestock who are grazing on pasture have the best likelihood of survival. When eating grass, a number of livestock grazing on the ground took in toxic particles from fallout and polluted water. Because of the radioactivity of fallout particles, they could cause serious damage in their digestion systems. Since fallout particles could produce only 140 R of gamma radiation in just a few days, combination of external radiation, beta burns, as well as internal radiation are expected to cause death to the majority of animals that graze.

Medications to be taken when eating meat

In locations where the amount of fallout was insufficient to cause sickness, animals meat is safe to consume. Animals' internal organs could be enriched by radioactive molecules and atoms when they consume or drink water or food that has been contaminated with fallout. Avoid foods that contain thyroid, kidneys or liver. It is recommended not to eat any animal that appears to be sick or in some way. Due to severe or even fatal radiation exposure it is likely that the animal has acquired an infection with bacteria. Even if the beef is cooked thoroughly there are some bacteria that can produce toxic substances that can cause serious illnesses and even cause death. Every beef cut should be cooked to the point that it is no longer able to retain its pink color as a protective measure in times of crises. To get the best results it is recommended that the meat be cut into 1-inch thick slices prior to cooking. This preventative action, in addition to decreasing time of cooking and fuel consumption can also be beneficial.

In the wake of the disaster massive swaths across the United States would be spared the slaughter of animals that graze. There will be enough to eat and breed among those millions of animal species who had survived to allow a nation to be rebuilt. Radiation doses that exceed 100 millisieverts (mSv) rarely cause permanent loss of fertility. There are numerous animal studies suggesting that radiation-induced animals could produce healthy offspring.

The nations with the highest population depend on the production of grains, beans and vegetables to maintain large populations, even though the majority of livestock that produce food are killed. This means that the vast majority. It is quite difficult for those who live in areas that are not food producing regions to obtain the non-cooked food items that are that are stored in these places. People who have survived famines will be amazed to discover how little transportation is required to supply sufficient raw food. An enormous nuclear attack will not affect the food-carrying truck fleet as well as the fuel they require." The strength of American management and morale will enable millions of people living in poverty regions to receive food that is basic and unprocessed within the first few weeks. Due to the presence of cesium and strontium, radioactive elements within the earth, those who consume food products after an attack on nuclear power tend to develop cancer. Within 30 years any additional cancer deaths that result from radiation from outside would render the increase of this magnitude insignificant. In the event of an imminent nuclear war, cancer deaths aren't enough to endanger humanity's existence. Most of the people who survive large-scale attacks will be in areas in which they'll have a difficult time fending for themselves. A survival strategy shouldn't rely on the wild animals and plants as sources of food.

Food for homeless people

Most people could survive on tiny amount of food for several weeks following a nuclear explosion or fallout. However, after those who survived the fallout and blast leave their shelters, they'll have to endure a long period of suffering and deprivation. In the end the residents of shelters should be able to eat balanced and nutritious meals for several weeks in order to ensure their mental and physical well-being. A little less than a week's worth of food items are stored in the majority of American households. The ability to have a two-week supply of food that is easily transportable in the pantry is a significant benefit to survival. Residents of shelters would need to make difficult choices regarding what they should eat every day if they didn't be aware of when they could get more. Vitamins and other nutrients required to ensure a balanced diet won't be the main concern for people who have been well-fed. The short-term energy needs can be met through eating healthy food items that are high in calories, but without fat. In the event of water shortages and high protein meals such as meat are best consumed in moderation because they require more water than diets high in carbs.

Benefits of Getting Nutrients

* The liver reserves in vitamin A can be enough to safeguard the human body from vitamin A deficiency for a long time regardless of whether their diet is lacking in this essential nutritional element. Vitamin A deficiencies affect children faster than adults. Inability to see clearly in dim lighting is the first indication. The body's tissues change because of the long-term deficiency. Vitamin A deficiencies in newborns as well as children could result in growth slowing and even blindness. Vitamin A is usually present in dairy items, such as butter, milk and margarine. None of which would be easily accessible to all survivors.

* The cause of scleroderma is an insufficient or deficiency of Vitamin C. People who consume only grains and beans , and who are aware of the best way to grow them, and create sufficient vitamin C could be diagnosed with a fatal illness at first. After just four or six weeks of eating a diet high with vitamin C. Following the procedure bleeding and heart failure could cause death. One kilo of the crystallized

"ascorbic acid" form of vitamin C pure is the most effective and least costly method to safeguard yourself, your familymembers, and your friends from Scurvy. In contrast to tablets of vitamin C, the potency of vitamin C crystals is constant over time. Consuming sprouted seeds can help treat or prevent the scurvy problem, and not just sprouts. Scurvy was prevented during the Indian food shortage due to the growth of beans. Captain James Cook's crew relied upon an unfermented drink made of barley that was sprouting dry during the course of three years. To keep scurvy at bay during the long winters in northern China The Chinese have for a long time relied upon sprouting their beans. The risk of developing scurvy is reduced by taking a daily dose of 10 milligrams vitamin C. You can get 8-12 milligrams of vitamin C through the sprouting of a tiny amount of dry wheat or beans. The sprouting process can take up to up to 48 hours at the room temperatures. A minimum of two seconds of boiling in the water needed to avoid sickness and encourage the digestion of beans sprouted. The majority of vitamin C is removed from food items after cooking for a long period of time. Regarding Vitamin C's production common methods of sprouting produce more sprouts than is necessary. When using these techniques, clean water and numerous washes per day are necessary. Anyone, even those who are who are not shelters, could have to run out of water. After getting rid of garbage and damaged seeds and seeds, they are now in good condition to plant. A jar or plastic bag like one like a Ziploc is used to spread seeds into an inch-deep layer after they have been removed from water. With a plastic bag you'll need to make two large, loose sheets of paper, soak them in water before putting them in the bag's edge in the case that this is the method you prefer. Once seeds are sprouted Niacin, riboflavin as well as folic acid concentrations rise. When cooked, sprouting beans are more digestible; however, they are, however, less digestible and nutritionally lacking when compared to sprouted beans which were freshly prepared. Sprouts aren't enough. When seeds are sprouted in this way, the calorie content could be reduced. This means that the seeds that are sprouted lose some of their energy content.

* If there is no vitamin D calcium absorption becomes reduced. Therefore, rickets could be prevalent among children (a disorder of deficient calcium mineralization). A majority of Americans would be disconnected from their primary source from Vitamin D and fortified milk should a major nuclear explosion was to occur. Exposure to sun's UV radiation could trigger vitamin D within the body. Because of such an attack, more UV radiation can be absorbed by the Earth's surface, which could cause skin burns across the United States that is as serious as it is on the equator in today's time. The best place to get the maximum exposure to sunlight in cold weather is a tiny trench that is protected by the winds. Exposure to gamma radiation fallout particles within a narrow trench will provide about 90 percent protection.

* The recommended emergency fat amount includes 30 grams cooking oil or fat per day. This means that the emergency diet will not

need a greater amount of fat because fats are at a minimum following the nuclear attack. These calories are very small in comparison to the standard American diet, where fat is responsible for 40% of calories.

The majority of people do not consume iron supplements throughout their lives. Iron supplementation is vital for women who are breastfeeding or pregnant and toddlers. Cooking foods that are acidic like tomatoes in iron cookware as well as pans have increased the amount of iron available. Alternately, soak iron nails with vinegar for a few days until they start to rise. It usually requires between 2 and 4 weeks in order to finish the process. A solution of iron and vinegar that contains 30-60 milligrams of iron in a teaspoon would be sufficient for daily consumption. 10 milligrams daily is a recommended emergency dose. Incorporating iron nails into fruit, like an apple, can increase the iron content of food by a few days.

Food Storage

* Sugar and wheat can be preserved for a long time dry beans, nonfat milk powder and vegetable oil may be kept for years every year. Here are some storage guidelines

* Food must be maintained at all times dry. Drums and storage containers made of metal that have covers make the best containers for storing dry grains to ensure it stays dry. 5-gallon cans are easily transported by a vehicle due to of their light weight when it comes to evacuating. It is apparent that grain is dry, particularly in areas with high humidity often isn't dry enough for storage for a prolonged period of time. Drying agents make sure that the grain is dried enough to last for years. In this scenario the silica gel that has colour indication is a good choice as an effective drying agent. If it is able to absorb moisture, the solution will be blue. However, when heated, it changes to pink and is a great drying agent. Chemical supply firms in the majority of cities will sell silica gel at a fair cost. Silica gel can be used for many years by heating it in a hot oven or placed in the flame until it becomes blue

once more. Fabric envelopes made from fabric that are large enough to accommodate a cup of silica gel make most suitable containers for drying grain or determining dryness. It is suggested that a transparent window is sewn into the fabric to ensure that color changes can be seen. Fill the 5-gallon can up with grain up to a few inches of the top and then put in one envelope filled with silica gel. Then, seal the can close. Any lid that has fallen off can be secured by some pieces of tape. The grain is dry enough when the silica gel is blue after a couple of days. If the silica gel turns pink, repeat the procedure with fresh envelopes, until the grains can be seen as dry.

Chapter 9: The Light And Medicines Chapter

Tragedies have proven that, even in complete darkness, people are able to remain in peace for long periods of time. In contrast, others in a shelter filled with scared people may collapse into pieces if they don't discern anything and cannot leave the shelter. Imagine the impact of a handful of panicked people on the other inhabitants of the darkened shelter isn't difficult. While fighting, even the tiniest of light can discern between a calming situation and one that could be life-threatening. The 3rd night during their stay they Utah family witnessed an incident.

The flashlights belonging to the family as well as other electrical lights were gone from use. There weren't any candles in the house and the items listed that were on the Evacuation Checklist (cooking oil glass jar and cotton thread) were all gone. They could have managed to build a basic lamp from these items and keep a lighting going for weeks if they needed to. The mother's first experience with anxiety was in the early hours of two a.m. in the middle of her third day although she was a sane woman who was not terrified of darkness. The entire room was awoken by her declaration: "I need to leave this location. I'm totally lost."

Electric Lights

Electric lighting that relies on power from the public grid will likely be out of service outside of the fire, explosion and fallout areas. In addition to radiation from nuclear explosions' electromagnetic impact loss of power stations and transmission lines could cause the loss of the majority of electricity that is available to the public. The equipment provided at shelters for the public do not contain emergency lighting. Most likely candles and flashlights carried by some to shelters aren't going to last longer than a couple of days at the most basic level of light. A fantastic low-level source for continuous light is a low-amperage light bulb that is connected to a massive dry cell battery or car battery. A single dim 12-volt light bulb located in the dashboard of a car can last for 10 or fifteen hours.

Candles

If you're going for a shelter make sure that you have enough matches and candles stored in a water-resistant container such as a Mason Jar. If the shelter is overflowing and the humidity is high, it could be when matches that aren't stored in containers that are moisture-proof can't be lit in a single day. The best way to prevent gas buildup that can cause headaches by placing candles that are lit or other sources of fire near an air vent which lets fresh air into the shelter. To ensure that smoke is not emitted from homes that are burning nearby All candles and fires within the shelter must be smothered if the shelter is completely closed. Kerosene or gas-powered lights in a shelter is not permitted. The vapors they release can cause nausea, headaches or even cause death. These consequences can be serious in the event that gasoline lights or kerosene are knocked down through blast winds that surge into shelters that cover large areas.

Medicines

It is possible that Americans find themselves spending the duration of a week or more huddled in basements or shelters for emergencies due to the consequences. In this scenario it is important to take steps to stop spreading infectious diseases which includes both the regular precautions as well as new precautions. If medical facilities were not available and the importance of disease prevention would be much higher for us all. These strategies for preventing infections are simple to implement, however they require some control. The author has experienced and applied them in different desert, jungle and mountain settings while traveling and as an army soldier.

In addition, I've used similar methods when evaluating the survival capabilities of nuclear disasters across a variety of states. The provision of basic first aid could be much more crucial in the event of a major conflict, or war. As the vast majority of people have access to top first aid guidelines and manuals so we will not go over the fundamentals again.

Skin Conditions Prevention

Skin conditions are more common in shelters that are crowded, especially in summer, than is usually recognized. Ninety nine soldiers were housed inside an underground bunker that was cooled solely by summer air for twelve weeks during the Navy study. Although medical treatment for regular patients was available but there was a substantial incidence of skin problems. 560 sick calls were reported and 40 of the 100 healthy males suffered from the heat rash, while 23 also had skin issues like fungal infections, each of which was identified as sick calls. The reason for this is that these sailors were living in a poorly ventilated environment and did not clean their skins with sweat or use any of the preventive measures suggested below. In hot temperatures, even in ventilated shelters that are surrounded by air, skin conditions are a significant problem if specific hygiene measures are observed. Skin conditions are more likely to be experienced in humid, hot conditions. Since the beginning of time the inhabitants of the jungle have taken numerous precautions to prevent skin problems.

Dead skin and sweat should be eliminated. (Dead skin is constantly rubbing and flails off during humid and hot weather, which starts the process of decay.) Cleaning your body many times throughout the daily, is typical for forest dwellers. In humid conditions the excessive use of soap could result in skin irritation due to the rapid loss of natural oils for your skin. Cleansing your face with water and then wiping it gently with a towel can ensure that your skin is clear. A 6-inch bedsheet material is ideal. Wash on a plastic sheet with its edges slightly elevated so that the water that is filthy will be able to be cleaned in the future. (Use thin sticks or planks to hold your edges.) With an washcloth, gently massage your abdomen, armpits, neck and buttocks. and feet using approximately two-thirds of your precious water. Cleanse your hands in the remaining water then use them to wash your face. With just a few minutes hot water, clean your washcloths regularly on a daily basis.

To ensure maximum dryness of your skin, make sure you to ensure the hottest and most naked position If it's possible. It is best not to be sharing your bed with anyone apart from your immediate family members. Utilizing a strong solution of chlorine and cleaning your toilet seats is the ideal method to stop the spreading of germs. Socks and underwear must be disinfected or washed whenever possible. In order to ensure the health of those who live in the shelter, it's essential to disinfect clothes rather than washing them. The clothes can be disinfected by taking it to the sink and soaking in hot water. Avoid using chlorine bleach to wash clothes unless there is plenty of water to wash. Always wear closed-toe sandals or shoes to protect your feet from fungal illnesses.

Respiratory Diseases Prevention

Snorting and coughing in highly used shelters can trigger respiratory illnesses. The prevention of illness could be improved by sufficient ventilation. Patients who are coughing or sneezing are advised to stay close to the air exhaust vents in shelters that are small. Large shelters that have a lot of residents have a higher risk of having one or more being infected with a contagious disease in comparison to smaller shelters.

Radioactive Iodine is the cause of thyroid problems.

The radiation from nuclear reactors cannot be stopped from causing damage to cells within our body. However, potassium and iodine-based dement salt. A period of one hour to a day before radioactive iodine can be inhaled or ingested at all, even in very small quantities. About 99 percent of damage to the thyroid gland is prevented as a result of this treatment. Radiation- and non-radioactive-iodine-absorption by the thyroid gland is straightforward.

In the majority of instances it is located in either or both forms. In both cases, the thyroid gland absorbs regular, non-radioactive Iodine prior to any radioactive iodine can be made available to the blood. Iodine that is non-radioactive is absorbed and stored by the gland until it reaches saturation. In addition to radioactive forms that could end up in bloodstream, the thyroid is able to only absorb around 1 l (as more iodine than it absorbs). (The kidneys are quick to eliminate excess iodine from blood.) The thyroid gland is able to be able to handle a tiny amount of additional iodine with no damaging it. In the end small amounts of radioactive iodine present in the thyroid could cause abnormal thyroid cells because of the radiation levels they experience. The thyroid dysfunction can be one of them thyroid nodules as well as thyroid cancer.

Conclusion

The thought of a nuclear attack on any country is devastation The death toll could be in the thousands and as many will be affected for many years to come due to radiation's harmful effects.

The nuclear weapons that are available today are far stronger than the ones used at the close in World War II; the consequences could be catastrophic.

But, it's not all doom and gloom.

While life as we are on Earth will change dramatically however, it is still likely to be able to survive the explosion and following radiation exposure.

Your chance of survival is to begin preparing your body today. The first blast can cause heat-related death, or the shockwave, or even make you blind from its intensity. It is important to take precautions against this.

The other problem, which could be a few minutes after the blast, contingent on wind strength and direction is radiation. To shield yourself from radiation it is essential to stay inside the underground structure, with strong walls of concrete or earth. Stone can also be very effective.

Making a grocery list doesn't need to be costly or time-consuming. You just need to come up with an outline. By adding a few more products to your daily shopping list can quickly create a large store.

Earth bags can be bought at a cost of PS150 per bag. creating the shelter can be done by you, just a bit at one time. The only thing that stands in the way is the amount of time you'll have until the moment that the first missile hits and preparing now is vital.

www.ingramcontent.com/pod-product-compliance
Lightning Source LLC
Chambersburg PA
CBHW050407120526
44590CB00015B/1863